WLAN Positioning Systems

Describing the relevant detection and estimation theory, this detailed guide provides the background knowledge needed to tackle the design of practical WLAN positioning systems. It sets out key system-level challenges and design considerations in increasing positioning accuracy and reducing computational complexity, examines design tradeoffs, and presents experimental results.

Radio characteristics in real environments are discussed, as are the theoretical aspects of non-parametric statistical tools appropriate for modeling radio signals, statistical estimation techniques, and the model-based stochastic estimators often used for positioning. A historical account of positioning systems is also included, giving graduate students, researchers, and practitioners alike the perspective needed to understand the benefits and potential applications of WLAN positioning.

Azadeh Kushki is a Postdoctoral Fellow at the Bloorview Research Institute, Holland Bloorview Kids Rehabilitation Hospital. She received her Ph.D. in Electrical Engineering from the University of Toronto in 2008 and is a principal author of many articles on the topic of WLAN positioning.

Konstantinos Plataniotis is a Professor at the University of Toronto, where he directs the Multimedia Laboratory. He is also an Adjunct Professor with the School of Computer Science at Ryerson University. He has contributed chapters to 15 books, co-authored the book *Color Image Processing and Applications* (2000), co-edited *Color Imaging: Methods and Applications* (2006), and published more than 350 technical papers.

Anastasios Venetsanopoulos is a Professor of Electrical and Computer Engineering at Ryerson University and a Professor Emeritus at the University of Toronto. He has authored eight books, contributed chapters to 30 books, and published over 830 technical papers. He is a Fellow of the Engineering Institute of Canada, the IEEE, the Canadian Academy of Engineering, and the Royal Society of Canada.

WLAN Positioning Systems

Principles and Applications in Location-Based Services

AZADEH KUSHKI
Holland Bloorview Kids Rehabilitation Hospital

KONSTANTINOS N. PLATANIOTIS
University of Toronto

ANASTASIOS N. VENETSANOPOULOS
Ryerson University

CAMBRIDGE
UNIVERSITY PRESS

CAMBRIDGE
UNIVERSITY PRESS

Shaftesbury Road, Cambridge CB2 8EA, United Kingdom

One Liberty Plaza, 20th Floor, New York, NY 10006, USA

477 Williamstown Road, Port Melbourne, VIC 3207, Australia

314–321, 3rd Floor, Plot 3, Splendor Forum, Jasola District Centre, New Delhi – 110025, India

103 Penang Road, #05–06/07, Visioncrest Commercial, Singapore 238467

Cambridge University Press is part of Cambridge University Press & Assessment,
a department of the University of Cambridge.

We share the University's mission to contribute to society through the pursuit of
education, learning and research at the highest international levels of excellence.

www.cambridge.org
Information on this title: www.cambridge.org/9780521191852

© Cambridge University Press & Assessment 2012

First published 2012

A catalogue record for this publication is available from the British Library

Library of Congress Cataloging-in-Publication data
Kushki, Azadeh.
　WLAN positioning systems : principles and applications in location-based services /
　Azadeh Kushki, Konstantinos N. Plataniotis, and Anastasios N. Venetsanopoulos.
　　p.　cm.
　Includes bibliographical references and index.
　ISBN 978-0-521-19185-2 (hardback)
　1. Wireless communication systems–Technological innovations.　2. Mobile
　communication systems–Technological innovations.　3. Wireless LANs.
　4. Aids to navigation.　5. Global Positioning System.　I. Plataniotis, Konstantinos N.
　II. Venetsanopoulos, A. N. (Anastasios N.),1941–　III. Title.
　TK5103.2.K874　2012
　623.89´33–dc23　　　2011040515

ISBN　978-0-521-19185-2　Hardback

To my family for their never-ending love and support. A.K.

Contents

    3.1   The location stack   30
    3.2   Proximity detection   32
    3.3   Lateration   33
           3.3.1   Circular lateration   33
           3.3.2   Hyperbolic lateration   34
    3.4   Angulation   36
    3.5   Fingerprinting   37
    3.6   Dead reckoning   38
    3.7   Computer vision   38
    3.8   Comparison of positioning techniques   39
    3.9   Chapter summary   40

4   Positioning systems   **42**

    4.1   The Global Positioning System   42
    4.2   Cellular-based positioning systems   45
    4.3   Ultrasound and infrared systems   47
    4.4   Wireless local area network (WLAN) positioning   48
    4.5   Comparison of positioning systems   48
           4.5.1   Evaluation criteria   48
           4.5.2   Evaluation   50
    4.6   Chapter summary   51

Part II Signal processing theory

5   Positioning in wireless local area networks   **55**

    5.1   Wireless local area networks   55
    5.2   Radio signal features in WLANs   57
    5.3   Characteristics of the example environment   57
    5.4   Properties of received signal strength   59
           5.4.1   Spatial properties   59
           5.4.2   Temporal properties   61
    5.5   Modeling the RSS–position relationship   64
           5.5.1   Parametric modeling   64
           5.5.2   Fingerprinting-based methods   65
    5.6   Technical challenges in RSS-based positioning   66
    5.7   Chapter summary   67

6   Memoryless positioning   **68**

    6.1   The problem of statistical memoryless positioning   68
           6.1.1   Optimality criteria   69
           6.1.2   Statistical radio map model   71

Preface

For thousands of years, location information has been used for navigation. This has changed in the last century as advances in wireless communication and microelectronics have given birth to *mobile computing devices*. These devices enable their users to access sensing and computing capabilities from anywhere and at any time. An important consequence of such mobility is that the resource and information needs of wireless users are no longer fixed but vary with their changing location and, more generally, with their changing *context*. This has sparked a new generation of applications that employ location information to cater to the changing needs of mobile users. These applications, known as *location-based services* (LBS), are offered on top of wireless communication infrastructures to add value to existing services.

To enable and support the delivery of LBS, accurate, reliable, and realtime user location information is needed. This need has incited a new interest in positioning and tracking systems whose aim is to determine the physical coordinates of a wireless mobile device carried by a human user. The focus of this book is one class of positioning systems that employ radio signals from wireless local area networks (WLAN) for positioning. These systems are of special interest as they are able to provide high positioning accuracies in indoor and outdoor environments with minimal deployment and maintenance costs.

This book is divided into two parts. The first part focuses on topics related to the history and applications of positioning systems. The second part is dedicated to technical issues related to designing WLAN positioning systems.

Part I of the book begins by providing a brief history of positioning and navigation and the origins of radio-based positioning (Chapter 1). Next, we discuss location-based services offered in wireless networks and their applications (Chapter 2). We then present the fundamentals of positioning techniques and existing systems that enable the delivery of location-based services (Chapters 3 and 4). We examine the advantages and limitations of each positioning method and introduce WLAN-based positioning.

Part II of the book examines the technical issues involved in converting WLAN sensor readings into position estimates. We begin by discussing the nature of sensor measurements in WLAN positioning in Chapter 5 and the challenges involved in computing with these measurements. We then proceed to discuss the details of algorithms for position computation using WLAN signals in Chapter 6. In particular, we use the framework of statistical signal estimation and nonparametric methods to develop several position estimators.

Due to the noisy nature of WLAN radio signals, position estimates obtained from these signals are often noisy and associated with high uncertainty. To mitigate the adverse effects of these factors, we present two different methods. In Chapter 7, we discuss the use of pedestrian motion models in addition to radio signals to improve positioning accuracy and reliability. In particular, we examine the use of various adaptive filters for optimal fusion of information obtained from the motion model and radio signals. The second approach for combating noise and uncertainty associated with radio signals involves intelligent selection of sensors used for positioning. Chapter 8 is dedicated to discussing this topic.

Having the fundamental tools needed to address position estimation, Chapter 9 proceeds to discuss issues related to design and architecture of positioning systems. Finally, Chapter 10 concludes the book and provides directions for future research.

Part I

History and applications

1 Positioning through the ages

For thousands of years, the ability to explore the world has significantly impacted human civilization. Human explorations have enabled the interaction of cultures for the purposes of geographic expansion (for example, through war and colonization) and economic development through trade. These interactions have also played a pivotal role in an exchange of knowledge that has supported the advancement of science, the development of religion, and the flourishing of the arts throughout the world.

World exploration is largely enabled by the ability to control the movement of a vessel from one position to another. This process, known as navigation, requires the knowledge of the locations of the source and destination points. The process of determining the location of points in space is known as positioning. In this book, we use the terms location and position interchangeably to refer to the point in physical space occupied by a person or object.

Throughout history, various positioning methods have been developed including methods using the relation of a point to various reference points such as celestial bodies and the Earth's magnetic pole. More recently, the advent of wireless communications has led to the development of a number of additional positioning systems that enable not only navigation, but also the delivery of additional value-added services. The focus of this book is one such positioning method that employs wireless local area signals to determine the location of wireless devices.

In this chapter, we provide a brief account of the historical development of navigational techniques (Section 1.1). As shown in Figure 1.1, we consider two distinct periods. The first period, termed the Age of Traditional Navigation, refers to the development of navigational techniques developed to facilitate exploration and sea travel before the nineteenth century (Sections 1.2 and 1.3). The second period, termed the Age of Modern Navigation, begins with the advent of wireless communication, which ultimately gave rise to the positioning systems in commercial use today (Section 1.4).

1.1 Origins of navigation

The development of navigational science was necessitated by the human need to roam about the world. In ancient times, travel played an important role not only in exploration, but also in trade, conquest, and religious and cultural expansions. For example, colonization and the spread and development of the major religions of the world were made possible because of the human ability to move between distant locations. Much of

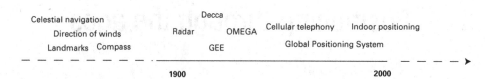

Figure 1.1.   Overview of the history of navigation.

the world remained unexplored and unmapped during ancient times. As such, effective positioning and wayfinding techniques were needed to guide travel between two points. Clearly, navigation requires knowledge of one's position as one travels along the path connecting the source and destination of travel.

Ancient civilization traveled by land and sea. At this time, navigation was made possible through the use of known landmarks to position oneself. The landmark-based mode of wayfinding was sufficient at the time when sea travel was used primarily for hunting and fishing. However, as time passed, the potential of sea travel for exploration was unlocked. Such exploration necessitated long-haul voyages that required sea vessels to travel great distances from the shoreline. This rendered the landmark-based mode of navigation insufficient, and motivated the development of more advanced positioning and navigation techniques [74].

1.2   The age of traditional navigation

Ancient navigational techniques generally obtained one's location relevant to reference points with known positions. In this section, we briefly review some of these techniques.

1.2.1   Navigation based on landmarks

As mentioned previously, known landmarks, such as the shoreline, were used to localize to determine the course of travel of vessels. In the case of sea travel, ships and boats used the shoreline as a reference for wayfinding during the day. For night travel, lighthouses were built to guide travelers along the coast. One of the seven wonders of the ancient world, the Pharos of Alexandria, is an example of such a lighthouse built in Egypt around 200 years before Christ.

In addition to the above methods, other modes of navigation in the ancient world employed techniques that relied on observing the direction of winds and sea currents to determine the direction of travel of vessels.

1.2.2   Celestial navigation

Ancient civilizations such as the Phoenicians, Greeks, Persians, Polynesians, Vikings, and the Chinese commonly used on a navigational technique known as *celestial navigation*. This technique employs celestial bodies, including the sun, the moon, and various navigational stars, as references for positioning. Crude modes of celestial positioning have been used for thousands of years. For example, in Homer's epic work, the *Odyssey* (dating back to approximately 1000 years before Christ), the hero Odysseus uses the Great Bear constellation to navigate.

Celestial navigation uses angular measurements between these celestial bodies and the horizon to obtain one's position. This required the development of instruments for measuring angular distances from the horizon. An example of such an instrument is the *astrolabe*. The invention of this instrument is attributed to the Greek astrologer and mathematician Hipparchos in the second century before Christ [74].

Celestial navigational techniques in the ancient world were crude and imprecise. These basic techniques were further refined and perfected over several centuries and led to the development of precise maritime navigation as we shall see in later sections. In fact, the basic ideas of the ancient world underlie many of the positioning systems used today [74].

1.2.3   The compass

Compass-based navigation is based on the use of the earth's magnetic pole as a reference. A compass consists of a magnetized needle that aligns itself with the earth's magnetic pole. As such, this device can be used to determine the heading or direction of vessels. Figure 1.2 shows an example of a modern compass. Though the origins of the compass date back several thousand years, the first reference to the use of this device for navigation is found in the twelfth century [74].

Figure 1.2.   Example of a modern compass. Image © iStockphoto.com/Ldf.

1.3   The age of exploration

As time passed, the aforementioned techniques were perfected to allow for long-distance travel both by land and sea. For example, the Middle Ages saw significant advancement in the science of navigation by Islamic and Persian scholars who improved measurement techniques used in celestial navigation and created detailed maps of the known world. For example, Islamic navigators invented the *kamal*, an instrument used for measuring the angle between the horizon and a navigational star and further improved existing navigational instruments such as the astrolabe. Figure 1.3 shows a photograph of a Persian astrolabe from the thirteenth century. The scholars of this period also contributed significantly to the development of cartography (map making) and geographical sciences.

Given this backdrop, the Age of Exploration began in the fifteenth century when European ships set out to conquer and explore new lands. It was during the Age of Exploration that Portuguese and Spanish explorers including Christopher Columbus (fifteenth century), provided accounts from distant lands that led to the interaction of

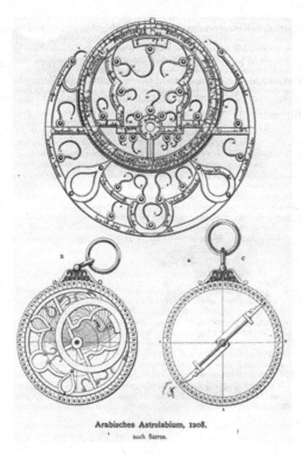

Arabisches Astrolabium, 1908,
nach Sarrus.

Figure 1.3.   A Persian astrolabe from the thirteenth century. "Astrolabe," Wikipedia, The Free Encyclopedia, http://en.wikipedia.org/wiki/Astrolabe.

Figure 1.4.   A German astrolabe from the sixteenth century. "Astrolabe," Wikipedia, The Free Encyclopedia, http://en.wikipedia.org/wiki/Astrolabe.

the Old and New Worlds. The subsequent two centuries marked a significant milestone in the development of systematic methods for navigation that enabled long-haul travel between the continents. During this time, traditional navigational charts and instruments were further refined (see, for example, a German astrolabe from the sixteenth century in Figure 1.4).

By this time, a well-established coordinate system was in use that represented the position of each point on earth using two pieces of information, namely *latitude* and *longitude*. Figure 1.5 depicts this coordinate system.

Latitude is the angular distance north or south of the equator (the equator has a longitude of 0 degrees, whereas the north and south poles have latitudes of 90 degrees north and south, respectively). Latitude was obtained using instruments such as the astrolabe which allowed the measurement of the angular distance between two objects. Near the end of the seventeenth century, Sir Isaac Newton invented the quadrant, used for the measurement of angular distances between two objects. This paved the way for the development of the octant and, consequently, the sextant, further perfecting the measurement of latitude. Figures 1.6 and 1.7 show examples of these instruments.

Longitude is the angular distance between a point's meridian and the Prime Meridian. One way to compute longitude involves calculating the time difference between one's position and a fixed point (because the earth rotates 15 degrees in one hour, knowledge of the time difference between the two points provides an angular distance). As such, measurement of longitude requires precise knowledge of time. However, since this knowledge was not available to early navigators, they had to rely on methods such as dead reckoning to find their longitude. This led to inaccurate navigation, which resulted in prolonged voyages and misdirected ships. Tragically, errors in computation of longitude caused the Scilly naval disaster of 1707 when almost 2000 lives were lost.

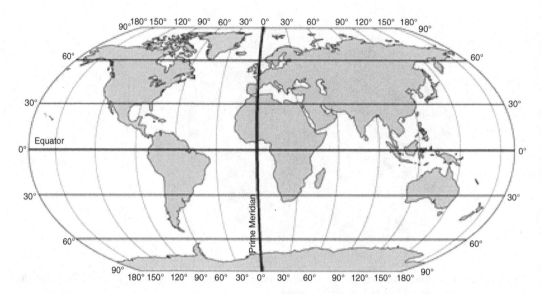

Figure 1.5.   Latitude and longitude lines. Horizontal and vertical lines correspond to latitude and longitude lines, respectively.

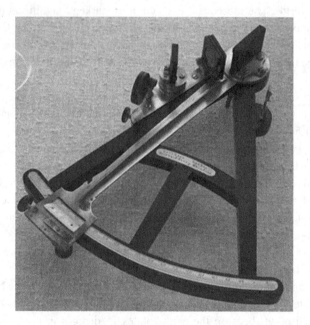

Figure 1.6.   An octant. "Octant", Wikipedia, The Free Encylopedia, http://en.wikipedia.org/wiki/Octant.

This problem of longitude was so significant to navigators of the seventeenth and eighteenth centuries that several organizations, including the British, French, and Spanish governments, offered prizes for finding the solution to the longitude problem. This led to the development of the marine chronometer, initially proposed by John Harrison. This

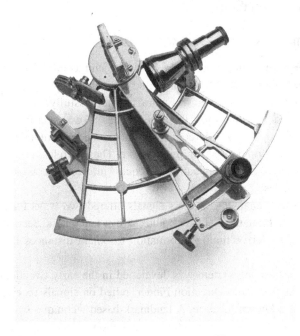

Figure 1.7. A sextant. Image © iStockphoto.com/DNY59.

Figure 1.8. John Harrison's marine chronometer. "Marine chronometer," Wikipedia, The Free Encyclopedia, http://en.wikipedia.org/wiki/Marine_chronometer.

chronometer is shown in Figure 1.8. The device allowed for precise measurement of time at a fixed location, which in turn enabled sailors to measure longitude given the local time.

The eighteenth and nineteenth centuries further saw the refinement of tools used for measurement of latitude and longitude. These remained the primary modes of navigation until the dawn of modern navigation in the twentieth century.

1.4   The age of modern navigation

1.4.1   Radio-based systems

The advent of wireless communications in the late nineteenth century marked another important step in the development of positioning and navigation systems. On December 12, 1901, Guglielmo Marconi (pictured in Figure 1.9) successfully accomplished the first transatlantic wireless transmission – the Morse code "S" was sent from Cornwall, England, to Newfoundland, Canada. Marconi also demonstrated the potential of wireless telegraphy for communication between ships and shore. These wireless communication systems were initially used in sea navigation to request information, assistance, or to transmit warning messages.

As we shall see in later chapters, wireless signals offered two ways for improving positioning systems [74]. First, these signals could serve as landmarks. Second, wireless signals could be used to derive timing information to obtain distances from known positions.

The first wireless positioning system was developed in the early twentieth century. This system, known as the Radio Direction Finder, relied on signals received from a radio transmitter with a known location. A landmark-based technique was employed whereby a directional antenna was used to find the direction of incoming radio signals, and hence that of the known landmark.

Figure 1.9.   Guglielmo Marconi. "Marconi," Wikipedia, The Free Encyclopedia, http://en.wikipedia.org/wiki/Marconi.

In the early twentieth century, the first prototypes of the RAdio Detection And Ranging (radar) system were developed. The radar system transmits pulses of radio waves and measures the reflected waves to determine the altitude, bearing, range, and speed of objects. Subsequently, several other radio-based positioning systems were developed including the GEE, LOng RAnge Navigation (LORAN), OMEGA, and Decca systems. These systems belong to the class of hyperbolic positioning systems which use the time difference between the arrival of radio signals from two transmitters to compute the location of a receiver (further details on this method are provided in Chapter 3).

Another important development in recent years is cellular telephony. The primary application of cellular systems is mobile communication. However, in 1996, the United States Federal Communications Commission introduced the E-911 mandate which requires mobile operators to know the position of their users within prescribed limits. This led to the development of cellular-based positioning systems.

1.4.2   Satellite-based systems

In 1951, the first artificial satellite, Sputnik, was launched into the earth's orbit. The success of Sputnik precipitated the space race which in turn led to significant advances in space travel. Though the intended application for artificial satellites was not positioning, it was soon realized that the position of a satellite could be accurately determined by observing its transmissions. The promise of this technique led to the development of satellite-based positioning systems in the late twentieth century, most notably the Global Positioning System (GPS), which was made available for civilian use in the 1990s. This system is explained in detail in Chapter 3.

The development of satellite-based positioning systems, together with the advent of mobile computing, have allowed mass market, commercial use of positioning information in a new generation of applications known as location-based services. These services have revolutionized the way we consume positioning information. Chapter 2 is dedicated to a detailed discussion of these services.

1.5   Chapter summary

The need for location information is a direct result of the human need for exploration, trade, and conquest. The origins of navigational sciences date back to thousands of years before Christ. Until the nineteenth century, location information was obtained mainly by employing celestial bodies as reference points. This changed with the advent of wireless communication, which allowed propagation properties of wireless signals to be exploited to obtain location information. A prominent example of wireless positioning systems is the Global Positioning System (GPS), which employs artificial satellites as reference points for positioning. The tremendous success of the GPS system in civilian applications, together with the maturation of mobile computing, have kindled a new era in the history of positioning systems and the dawn of a revolution in the way location information is generated and consumed.

2 Location-based services

Traditionally, the application scope of positioning systems was limited to target tracking and navigation in civilian and military applications. This has changed in past decades with the advent of mobile computing. In particular, the maturation of wireless communication and advances in microelectronics have given birth to *mobile computing devices*, such as laptops and smart phones, which are equipped with sensing and computing capabilities. The mobility of these computing devices in wireless networks means that users' communication, resource, and information needs now change with their physical location. More specifically, location information is now part of the *context* in which users access and consume wireless services. This, together with the availability of positioning information (for example, through the Global Positioning System), both necessitated and enabled the development of services that cater to the changing needs of mobile users [34]. This need has sparked a new generation of applications for positioning known as *location-based services* (LBS) or *location-aware systems*. Formally, LBS have been defined in many ways [24, 50, 80]. In this book, the term LBS is used to indicate services that use the position of a user to add value to a service [50].

In this chapter, we will discuss the economical and ethical implications of LBS. We begin with an assessment of the market potential for these services (Section 2.1). This is followed by a discussion of application areas where LBS services can be employed (Section 2.2). Finally, we discuss the ethical implications of LBS (Section 2.3).

2.1 Market potential

The mass market proliferation of hand-held devices, such as smart phones and tablet computers, has created an attractive market for services that cater to the changing needs of the mobile user. Location-based services (LBS) are examples of such services. LBS are especially attractive because they can be offered without the need for any additional hardware cost and because they can be delivered on top of existing communication infrastructures. The 1990s saw the inception of various LBS targeted towards users of mobile devices in personal and commercial settings. Over the next few years, the world market for LBS is projected to grow significantly, reaching 10 billion dollars in 2014 (see Figure 2.1).

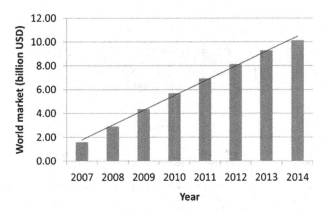

Figure 2.1. Projected market for LBS and software. (Data source: Philip M. Parker, INSEAD, www.icongrouponline.com.)

In addition to the rapid growth of the LBS industry, there are several factors that make this an attractive market to enter. This section analyzes the market potential and attractiveness of the LBS industry by evaluating opportunities and challenges in a competitive environment. To accomplish this, we use the framework of Porter's five forces. These forces consist of five factors that affect a company's profitability in a given industry. These include the power of buyers and suppliers, and threats from established rivals, new entrants, and substitute products. Figure 2.2 provides an overview of this analysis.

2.1.1 Industry analysis

In this section we analyze the LBS industry. In particular, we show that the "producers" of LBS are companies that provide software applications that employ existing communication infrastructures to deliver a value-added service. The "consumers" of LBS are users of mobile devices such as laptops, tablet computers, and mobile phones. While LBS can be offered to both consumer and business markets, the analysis of this section focuses mainly on the former. LBS can be distributed by mobile providers (for example, the E-911 service) or they can be distributed directly to the mobile users as application downloads (for example, the Apple App Store).

Buyer power

Buyer power refers to the influence that consumers can exert on a given industry. For example, if there a few concentrated buyers with significant market share in a given industry, these buyers will have the power to influence the price and quality of offered products. For the LBS industry, we assume that the buyers are consumers of LBS (that is, mobile users).

One factor that affects buyer power is the volume and concentration of buyers. In the case of LBS, there is a large volume of fragmented buyers. In fact, currently, the number

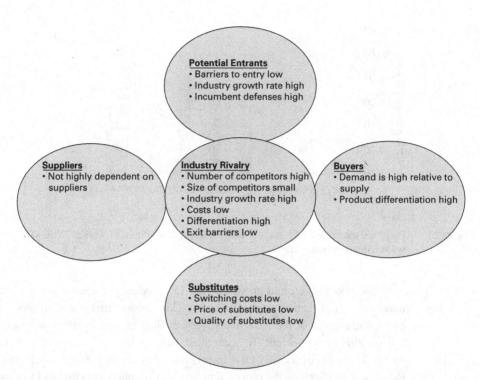

Figure 2.2.   LBS market analysis using Porter's five forces.

of users of LBS (demand) is much higher than the number of LBS providers (supply). According to a report published by Simba Information, an estimated 86% of Americans over the age of 13 own a mobile device. Simba Information further estimates a steep growth in the number of users of LBS worldwide, which is projected to rise to nearly 500 million in 2012 (see Figure 2.3). These numbers suggest a large volume of buyers for LBS. Moreover, buyers of LBS in this discussion are personal users of mobile devices. Therefore, buyers are generally fragmented with low bargaining leverage. The fact that the buyers of LBS are fragmented (many different buyers), suggests that no particular buyer has significant power over the product or price.

Another factor that can affect buyer power is their availability to switch to a similar or substitute product. As we shall see in this chapter, LBS can enrich many different application areas. As such, existing LBS products are generally highly differentiated. Note that as the number of players in the market increases, product differentiation will decrease.

Collectively, the above evidence suggests a low buyer power, indicating that suppliers of LBS have the power to set the price and market trend for these service offerings. However, caution must be taken in interpreting this evidence as buyer power is likely to increase with the growing number of LBS suppliers and decreasing product differentiation.

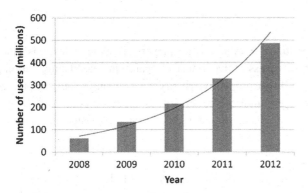

Figure 2.3. Projected number of LBS users worldwide. (Source: Simba Information.)

Supplier power

Suppliers refer to entities that provide the LBS industry (producers) with raw materials needed to develop and deliver their services. Powerful suppliers can significantly affect the pricing and product lines of the industry. For example, suppliers may raise the price of raw materials needed for manufacturing.

To understand supplier power in the LBS industry further, let us take a closer look at how LBS are developed and distributed. The development of LBS generally involves software design and implementation. As such, labor (and not raw materials) is the key resource required. Supplier power is weak in this regard. Because LBS providers must access software and hardware resources on mobile devices to develop LBS applications, another set of suppliers in the LBS industry are manufacturers of mobile device hardware and software. These suppliers, however, are also weak. This is due to a pressing need for product differentiation to increase competitive advantage in a highly saturated and competitive mobile device industry.

Once a location-based service is developed, it must be distributed to the consumers. Two distribution channels can be considered. First, users of mobile devices may download an LBS application onto their mobile phones and use built-in phone accessories to obtain positioning information needed for delivery of LBS (for example, GPS or WiFi receivers). This can be done through mobile software providers (for example, the Apple App Store) or directly through the LBS producers' websites (though this would result in increased cost). In either case, LBS can be distributed directly to the users and can operate independently of mobile telecommunication service suppliers.

The second distribution channel for LBS may be through mobile telecommunication service providers. In particular, these providers may "push" LBS to the subscribers (for example, advertising, E-911). Here, the delivery of LBS is contingent on a few powerful suppliers of telecommunication services, leading to strong supplier influence.

To summarize, supplier power in LBS is highly dependent on reliability of telecommunication services providers. A direct distribution channel can minimize this reliability and consequently, the supplier power.

Competitive rivalry

The degree of competitive rivalry among the players in a given industry is a major determinant of the competitiveness of that industry because increased competition generally leads to lower profitability. Competitive rivalry depends on various factors, such as the industry growth rate, the number and size of competitors, costs, differentiation, and exit barriers. Rivals can compete in many dimensions including price, innovation, and marketing.

Industry growth rate

A rapid industry growth rate is desirable as this leads to lower competition over market share. As seen from Figure 2.1, the worldwide market for LBS is rapidly growing. To investigate the industry growth further, we will use the Diffusion of Innovations theory, which models how a new technology spreads through a market. In particular, this model stipulates that a given technology is successively adopted by five groups of individuals. As shown in Figure 2.4, these groups are as follows.

- *Innovators.* Also known as enthusiasts, this group of individuals is the first group to adopt a technology.
- *Early adopters.* This group comprises those individuals with the highest degree of opinion leadership among the adopter categories and is the second group to adopt a technology.
- *Early majority.* This group takes a significantly longer time than the preceding two groups to adopt a technology.
- *Late majority.* This group adopts a technology only after an average number of people in the society have adopted the technology.
- *Laggards.* The last group to adopt a technology, this group comprises risk-averse individuals.

Based on the figures provided in Figure 2.4, we suggest that LBS technologies are currently in the late stages of adoption by the early adopters, or even in the early stages of adoption by the early majority. This suggests a significant market growth in this area over the next few years, suggesting low rivalry in this industry.

Number of competitors

A large number of competitors in a market increases competitive rivalry as competitors must compete for access to resources and consumers. Currently, the number of competitors in the LBS industry is low compared to the number of users. However, the rapid maturity of mobile devices and positioning systems, together with the increasing popularity of LBS, mean that the number of suppliers of LBS is likely to grow significantly in the coming years.

Costs

Higher costs negatively impact revenue and increase rivalry as competitors must increase sales volumes to mitigate low profit margins. One of the most attractive features of the LBS industry is the low cost involved in provision of these services. This is a consequence of the fact that LBS are offered as value-added services on top of existing

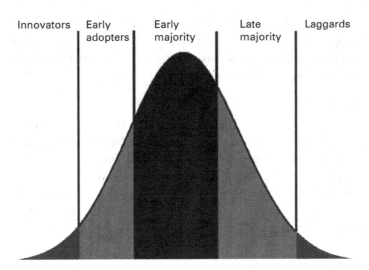

Figure 2.4.   The technology adoption cycle.

infrastructures. In its most simple form, LBS delivery incurs very low fixed costs. Consider, for example, LBS applications supplied through the Apple iPhone. This device is already equipped with GPS and WiFi capabilities. As such, obtaining location information requires interfacing with these devices through an available software development kit (SDK) provided by Apple. Moreover, once LBS software application is developed, it can be distributed to consumers at a small cost through on-line distribution channels such as the Apple App Store.

Size of competitors
Larger competitors have access to more resources that may enable them to produce higher quality products, achieve lower costs (for example, through economies of scale), and reach a larger number of consumers (for example, through effective advertising or distribution channels). As such, the size of competitors has a direct impact on competitive rivalry. As noted above, the cost and resources needed for development and delivery of LBS can be minimal. For this reason, the players in this industry are mainly small-size, startup companies (though larger size companies such as Google are now offering LBS). The size of these players will grow over time as the LBS market matures and the LBS applications become more technically sophisticated and require more resources. Moreover, the rapidly growing LBS industry and its attractiveness will encourage more, larger size, providers to enter this market. For example, telecommunication providers and mobile software and hardware manufacturers may enter the LBS market to capitalize on synergies and economies of scope.

Product differentiation
Low product differentiation reduces competitive rivalry as this provides LBS consumers with a wider range of providers from which they can choose. Combined with low switching rates, low product differentiation can significantly increase rivalry in the LBS industry.

Because the LBS industry is still in its growth stage, most product offerings target innovative applications. As such, product differentiation in this market is currently high. However, as the size of players in this industry grows, it may be possible that early, "maverick," products are replaced by high-quality substitutes.

Exit barriers

Exit barriers refer to the cost of abandoning a product or service when it is no longer profitable. Examples of factors that lead to higher exit barriers include fixed assets that depreciate quickly and cannot be easily sold, and long-term contracts. Low exit barriers increase competitive rivalry as they increase the willingness of competitors to enter the market. These barriers are generally low for the LBS industry. This is mainly due to the fact that fixed asset requirements are minimal. Moreover, long-term service contracts are not common-place for mass market, retail provision of these services.

Threat of entry

The threat of entry refers to the ease with which new players can enter the industry. Clearly, an increased threat of entry will result in increased competitive rivalry. The threat of entry for the LBS industry is high, due mainly to two factors. First, many providers of mobile technology are now providing software development kits (SDK) or open source code that can be used to build LBS applications with relative ease (consider, for example, the Apple iPhone and the Google Android). As a result, specialized knowledge is not generally required for entering this market. Second, as previously mentioned, the development of simple LBS applications generally incur a small cost, allowing small sized competitors to enter the market easily.

Substitutes

Substitutes are products in other industries that can potentially offer an alternative service that meets the same need. For example, paper maps are substitutes to GPS navigation. The existence of substitute products enables consumers to choose between providers, increasing competitive rivalry. In the case of LBS, the threat of substitutes is generally low as currently not many technologies offer similar services to those offered by LBS.

2.1.2   Assessment summary

The LBS industry is projected to grow significantly over the next few years both in terms of revenue and the number of users. The key advantages of the industry are the potential for significant growth phase, low supplier and buyer power, and low rivalry. Collectively, the above evidence suggests that the LBS industry is attractive.

The attractiveness of the LBS market will inevitably increase competitive rivalry by motivating new competitors to enter the market. Consequently, the industry attractiveness will diminish as the LBS and mobile computing markets mature.

2.2   Applications of location-based services

Historically, location information was primarily used for land and maritime navigation (see Chapter 1). By the end of the twentieth century, this information had also found wide applicability in aircraft navigation and target tracking in military settings. In these traditional applications, location information is consumed by users directly, in its *raw* form. For example, in navigation applications, the course of travel is adjusted by knowledge of the relationship between the vessel's current position and the destination.

The advent of mobile computing in the late twentieth century sparked a plethora of applications that revolutionized the way location information is produced and consumed. In particular, mobile computing devices enable users to generate, access, and consume information content *anywhere* and at *any time*. Because the information needs of mobile users change with their physical context, such information processing can be *personalized* by location. LBS are examples of such applications that exploit a mobile user's location to deliver context-relevant services. Figure 2.5 illustrates several application examples that may benefit from location personalization.

In contrast to traditional applications, in many LBS applications positioning information is not directly provided to the user. Instead, the user may receive *any* information content that is personalized based on location. For example, the user may receive location-relevant advertising or weather information on a mobile phone without actually receiving the raw location information.

Another important feature of LBS is that these services are offered on a number of geographic coverage scales. Positioning systems, such as GPS, allow provision of LBS

Figure 2.5.   Examples of LBS applications.

Table 2.1. Examples of indoor and outdoor LBS; note that the two categories are not mutually exclusive

Service	Indoor example	Outdoor example
Navigation	museums, hospitals	city travel
Emergency services	patient monitoring	E-911
Network management	security and authentication	hand-off and routing
Information delivery	forwarding services	geo-tagging
Social networking	personnel location	friend finder
Marketing	advertising in public spaces	restaurant advertising
Context-awareness	smart homes	recommender services

services on a global level that encompasses continents, countries, and cities (for example, roaming services and weather reports). At the same time, LBS can be offered on a finer scale comprising streets, buildings, and even rooms. As shown in Table 2.1, examples of indoor LBS include location-based network access, management, security, automatic resource assignment, context-awareness, health care, location-sensitive information delivery, and assisted navigation in public spaces such as shopping centers, museums, and hospitals.

Indoor LBS applications typically require positioning accuracies higher than that required in outdoor applications. For example, the Federal Communications Commission requires wireless providers to report location information with an accuracy of 50–300 meters depending on the technology used [23, 82]. Such accuracies will not be sufficient to deliver LBS on the scale of buildings and rooms. Moreover, the coverage of inexpensive and embedded GPS devices is limited in indoor environments [34, 82]. These limitations have motivated the development of positioning systems specifically geared towards indoor applications. Some of these methods will be reviewed in detail in Chapter 3.

The rest of this section reviews some of the well-established application areas of LBS in detail.

2.2.1   Navigation

The intended application of location information obtained from early radio- and satellite-based positioning systems was to aid military operations. This continued until the United States President Ronald Reagan made the Global Positioning System (GPS) available for civilian use. Initially, however, the signal quality available for civilian applications was deliberately degraded. This changed in 1996 when the United States President Bill Clinton declared GPS a dual-use system and disabled the selective availability feature of this system. Since then, GPS has been commonly used in aircraft and maritime navigation.

GPS receivers are portable and relatively inexpensive. Moreover, these are now available as dedicated GPS receiver devices (see Figure 2.6) or GPS receivers embedded on

Figure 2.6. Example of a GPS receiver used for navigation.

mobile devices (see Figure 2.7). As such, recent years have seen a mass market proliferation of these devices for consumer and personal applications. In particular, positioning information obtained from GPS is now used to assist navigation of cars and pedestrians. In both cases, navigation information can be augmented with on-demand maps and route-finding services. Many of these services also provide predictive information such as the time left to reach the destination.

In contrast to traditional navigation systems, where only positioning information is available to users, modern navigation systems provide their users with rich, non-location information such as traffic and weather conditions. Moreover, information about points of interest in the vicinity of a user can now be delivered either on-demand or in a proactive manner. For example, a list of restaurants together with their menus and relevant reviews may be provided to mobile users as they navigate through city streets. Google Maps is an example of such an LBS application that augments navigation with rich location-relevant information delivery on mobile devices. Figure 2.8 shows an example of a Google Map augmented with information about points of interest near the user's location as well as traffic and city transit information.

Whereas traditional navigation had been restricted to outdoor environments, the availability of fine-grade positioning information has now also enabled navigation in indoor spaces. Indoor navigation is used to guide people through unfamiliar spaces such as hospitals, museums, and shopping centers. This type of navigation can also be enriched with on-demand and personalized information delivery. For example, in a museum, relevant information regarding various relics can be provided based on the location of the person. Indoor navigation services can also be employed to guide individuals with disabilities through various spaces.

Figure 2.7.   Example of a mobile phone equipped with a GPS receiver. Image © iStockphoto.com/Alex Slobodkin.

2.2.2   Emergency and security services

One of the first LBS applications was the E-911 emergency service. Before cellular phones, emergency calls were always made from phones with a fixed location, allowing emergency personnel to locate callers in a timely and efficient manner. In the case of mobile phones, however, the location of the caller is no longer fixed and must be determined before the emergency response can take place. As such, one of the most important applications of LBS pertains to emergency services. These services can be offered when an emergency call is initiated by a mobile user (for example, a 911 call in the USA and Canada or 112 calls in Europe). These services can also be used to locate a mobile device for security purposes (for example, in the case of a kidnapping), for asset tracking and monitoring in the case of theft, and for search and rescue operations in the case of natural disasters.

Location-based emergency services can also be used to promote independence of vulnerable persons. For example, children can be tracked by their parents based on the location of their mobile device. Location-based emergency services may also enable hospital patients to roam around freely and be located in the case of a medical emergency. Furthermore, these services allow remote monitoring of at-risk outpatients and individuals with memory loss.

(a)

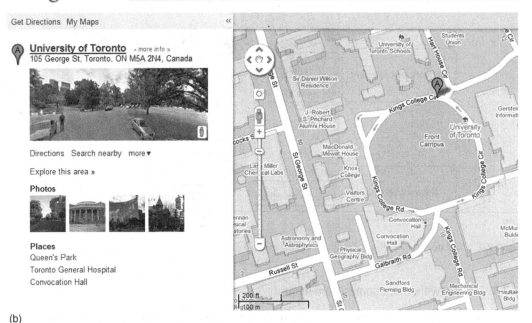

(b)

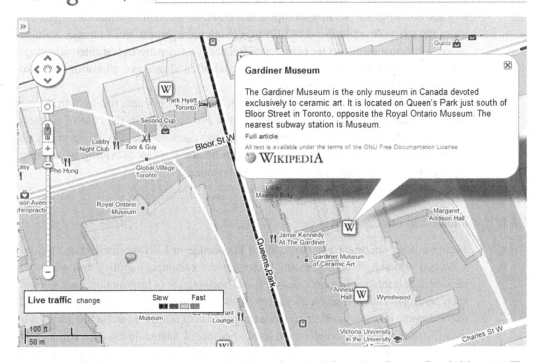

Figure 2.8. Examples of navigation enriched with non-location information. (Source: Google Maps.) (a) The map is augmented with information about places of interest near the user's location. (Map data ©2011 Google.) (b) The map is augmented with traffic and transit information as well as links to Wikipedia articles related to places of interest. (Map data ©2011 Google.)

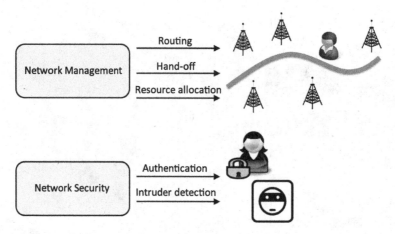

Figure 2.9.   Examples of LBS applications related to network management.

2.2.3   Network management and security

Because mobile devices can roam through large geographical spaces, communication and resource needs of these devices are dependent on the physical location of the device. As such, an important application area for LBS is network management, where communication functions are enabled and optimized based on location information. Figure 2.9 illustrates example applications of LBS in network management. For example, realtime location information allows for planning of frequency reuse patterns and antenna settings [24]. Additional functions, such as routing, hand-off, and resource allocation, can also benefit from location information [60].

Location information also enables location-based network security by introducing location-based authorization and authentication. For example, in the case of wireless local area networks (WLAN), where unauthorized users outside a building may gain access to wireless services, authentication can be achieved by providing networking access only to individuals inside the building boundaries.

2.2.4   Information management

Location is an important piece of context knowledge in LBS services related to *personalized* information management. These services encompass applications related to personalized generation, access, delivery, and organization of information. Examples of such personalized information access and delivery include location-based weather and traffic reports, information filtering (for example, lists of available restaurants and movie theaters), and information delivery related to points of interest in close proximity (for example, in tourism). Location-based reminders also provide users with personalized information when they are in prescribed zones.

Another example of location-aware information management pertains to the organization of *user-generated content*. The proliferation of mobile devices equipped with text and multimedia capabilities allows the production and sharing of mass volumes of information by users of mobile devices. Such information includes digital media (images, audio, and video) pertaining to social and personal events as well as blogs, podcasts, and reviews. Examples of notable repositories of user-generated content include Flickr (images), YouTube (video), and Facebook (social networking). Access to these mass volumes of information requires efficient organization and retrieval techniques that operate based on content features (for example, color content of images, or textual descriptor of videos). Additional meta-data pertaining to these media can further promote efficiency and effectiveness of access. Location information is an example of such meta-data as location is an important piece of context (for example, retrieve all photos taken of the CN Tower in Toronto). Augmenting user-generated content with location information is known as *geo-tagging*. Figure 2.10 shows an example of user "tweets" augmented with location information and location-based retrieval of images performed by Google Maps.

2.2.5 Social networking and entertainment

The widespread use of social networking applications has also given rise to LBS related to management of social contacts. These applications facilitate coordination of social activities such as meeting new people and organizing gatherings. These applications include friend finder and dating applications that broadcast user location information to friends' lists or alert users when a friend or interested party is in geographical proximity. Google Latitude (see Figure 2.11) is an example of such an application. Another example of location-based social networking is the Facebook Places application, which allows users to "check in" when they visit different places, to see friends who have checked into nearby places, and to "tag" friends based on location.

Positioning information is also employed to enhance collaborative gaming applications. For example, Geocatching is a collaborative treasure hunt game where participants around the world use GPS information to locate hidden treasure containers.

2.2.6 Marketing

Personalized marketing is key to reach consumers effectively. Since location is part of consumers' context, LBS can be used to deliver personalized advertising and promotions [71]. Marketing applications of LBS use positioning data to alert mobile users that a product or service, such as a restaurant, is in close proximity. Advertising applications of LBS provide a means for offering value to customers by delivering information when and where appropriate. Location-based marketing is also known as *geomarketing*.

Geotargeting is the method of determining the geolocation of a website visitor and delivering different content to that visitor based on his or her location, such as country, region/state, city, metro code/zip code, organization, IP address, or other criteria. A common usage of geotargeting is found in on-line advertising, as well as Internet television with sites such as iPlayer and Hulu restricting content to those geolocated in specific countries (also known as digital rights management).

(a)

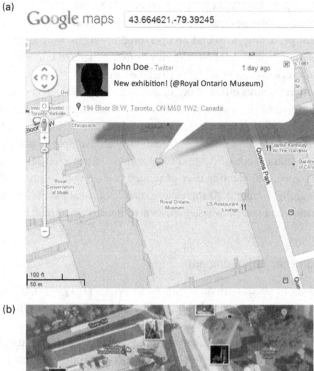

(b)

Figure 2.10.   Maps augmented with user-generated content. (Source: Google Maps.) (a) Map augmented with user "tweets"; (b) geo-tagging.

Mobile marketing often focuses on "push" interactive advertising. In this case, geo-targeting is used to determine the location of a user accessing a website and delivering location-relevant advertising. LBS advertising can direct users from wherever they are to a local merchant. Mobile devices equipped with positioning capabilities can also leverage location-based search opportunities. This opens up a new channel of "pull" search

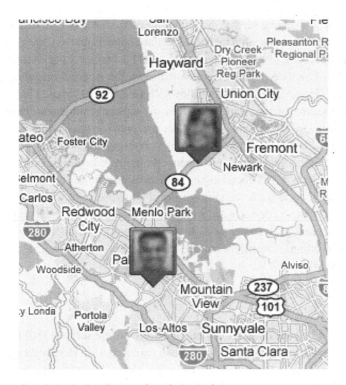

Figure 2.11.   Google Latitude. (Source: Google Latitude.)

marketing, where a mobile user's preference will determine the advertising material that is delivered.

There are significant opportunities for LBS marketing. According to a report published by Simba Information, the best prospects for interactive location-based marketing is through SMS, web browsers, and smartphone applications. This report also indicated that a significant percentage of mobile users were aware of mobile advertising and many were receptive to it.

- 60% of mobile data users found mobile advertising acceptable;
- 25% of mobile users recalled seeing advertising while using the web browser;
- 23% of mobile data users expected to see more mobile advertising;
- 32% were open to mobile ads if they lower their phone bill.

Many organizations have emerged that create content and marketing campaigns for mobile phones. Many of these are spin-offs of existing advertising agencies and web design/development companies. As such, there now exists a plethora of LBS.

- Facebook Places allows users to "check in" various places and receive special offers provided by nearby retailers. These offers can be shared with friends or used to earn reward points for repeat visits.

- Loopt is a social mapping service that offers location- and service-based advertising owned by CBS. The application includes localized banners on the CBS Mobile News and Mobile Sports sites which point users to local businesses based on their location.
- Visa now has a mobile payment service that uses the Android operating platform developed by Google. The application allows Visa card holders to use a locator feature to find the location of the nearest retailer offering items similar to those recently purchased on their card. Special deals are also presented on the phone as coupons that can be redeemed by showing the screen to the corresponding retailer.
- 3rd Dimension Inc., a mobile services network, partnered with the NBC to develop Mobileyes. Mobileyes is a location-based advertising application that includes live videos of weather and traffic conditions. Local advertisements are displayed as the updated traffic video is downloaded to the mobile device. A geotargeting capability is used to enable the advertiser to push an ad related to the direction in which the customer is heading. Advertisers include McDonald's, local auto dealers, and local tourism authorities.

2.2.7   Context awareness

LBS systems can be considered as subsets of *context-aware* services [50] that provide users with personalized services based on their changing context. Here context comprises information such as time, location, identity, and activity. The emerging application of Ambient Intelligence (AmI) is an example of efforts in context awareness. This term refers to environments with a multitude of embedded sensors that are aware of the presence of people and are responsive to their needs. These smart environments are integrated into daily lives to provide personalized services in areas such as surveillance, health, education, and security. Examples of functioning AmI environments include the Philips HomeLab project [1], MIT's Oxygen [15], and Georgia Tech's Aware Home [25].

2.3   Ethical considerations in location computing

The maturity of positioning systems (such as GPS) and the commercial success of LBS are two factors that have significantly contributed to the explosive growth in the amount of information pertaining to the location of mobile users. This has incited various concerns related to usage and storage of location information. In particular, since positioning involves direct monitoring of humans, it is imperative to ensure that usage of this information does not result in the infringement of civil liberties and rights [28]. To this end, before proceeding further, it is important to discuss the issues of consent and user privacy in positioning systems.

2.3.1   Consent

To facilitate the delivery of LBS, user location information is often disclosed to service providers. However, it is important that such disclosure is initiated and terminated with the consent of users. Moreover, it is pertinent that, prior to consent, users are informed of the exact implications of such disclosure. In other words, service providers must offer transparency with regards to implications of such disclosure. Examples of information that must be provided to users include the following.

- Who has access to the disclosed information? For example, will the service provider sell the information to marketing agencies?
- What steps, if any, are taken to protect the location information? These steps may include encryption of data.
- Will the information will be stored permanently by the provider and used in the future? This is specially important if the information is to be used to analyze patterns of user behavior for further personalization.
- What are the risks involved in disclosing the location information to the service provider? Such risks may include access by unauthorized parties, who may in turn disclose the information for profit, or maliciously disrupt the intended LBS.

2.3.2   Privacy

Location information can be used to infer personal information such as personal behavioral habits [8] and places of significance for users (for example, home address). As such, it is essential that positioning systems respect users' rights to privacy. This can be accomplished by minimizing the risk of unauthorized access to location information through secure computation, storage, and communication of any data pertaining to user location. One step toward this direction is promoting anonymity of user data. This may be achieved by using anonymous user identification tags as opposed to names. This method, however, is not a particularly effective means of privacy protection as user habits may be used to infer user identity [56].

2.4   Chapter summary

The mass market proliferation of mobile devices equipped with positioning capabilities has sparked a new generation of applications for positioning information. These applications, known as location-based services (LBS), offer added-value services to mobile users by personalizing information delivery based on location. In this chapter, we examined the market for location-based services and concluded that LBS are currently an attractive industry, primarily due to rapid growth and low costs. Furthermore, we discussed existing and emerging application areas for LBS. Lastly, we examined the ethical implications of LBS by outlining issues pertaining to consent and preservation of user privacy.

3   Positioning techniques

In the previous chapters, we discussed the history and application of modern positioning systems that enable the delivery of location-based services (LBS). In this chapter, we shift our attention to the fundamental positioning principles used in these systems. We begin this chapter by presenting the *location stack*, a model of location-aware systems, and identify the focus of this book (Section 3.1). We then proceed to discuss the most commonly used techniques for computing the position of mobile receivers. Similar to the techniques used in celestial navigation, modern positioning systems often employ a set of *references* with known locations for position computation. In this chapter, we discuss different positioning methods, differentiated by the type of references and signal measurements used (Sections 3.2 to 3.4). In addition to these techniques, which generally employ wireless signals, we will also briefly review dead reckoning (Section 3.6) and computer-based positioning (Section 3.7) methods. These two techniques employ modalities complementary to wireless measurements and as such provide a promising direction of development for hybrid positioning systems that employ multiple measurements to improve the accuracy and reliability of positioning. Finally, we conclude the chapter by discussing the advantages and disadvantages of each positioning method (Section 3.8).

3.1   The location stack

To position this book within the wealth of information available on positioning systems used in LBS, we review a model of location-aware systems proposed by Hightower *et al.* [35]. This model, known as the *location stack*, comprises seven layers, as listed below (see Figure 3.1).

(1) **Sensors**. This layer is responsible for the detection of raw data that can be used to obtain a location. The data detected vary according to the sensor modalities used. For example, the raw data reported in the GPS system are pseudo-range measurements, whereas the data used in computer vision-based systems are video streams.

(2) **Measurements**. This layer is responsible for transforming raw sensor data from the Sensors level into measurement types such as distance, angle, position, and proximity. This layer also provides an indication of uncertainty associated with the sensor that created the measurement.

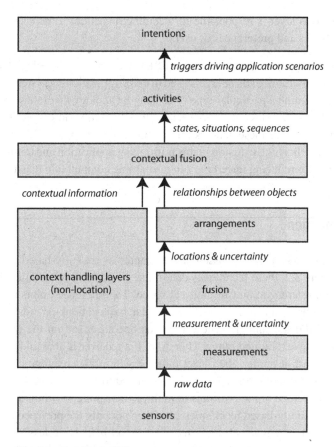

Figure 3.1.   The Location Stack [35].

(3) **Fusion**. Using the measurement information, this layer produces the position of objects (along with other auxiliary information such as acceleration or speed). This layer encompasses data fusion methods for continuously merging the measurements to produce a representation of the positions of objects of interest in the environment. The goal of this layer is to employ multiple measurements to reduce uncertainty associated with a computed position.

(4) **Arrangements**. This layer is responsible for inferring the spatial relationships between detected objects in the environment (for example, proximity and containment).

(5) **Contextual fusion**. This layer combines location data with non-location contextual information (for example, information obtained from calendars, email, and environmental sensors). The objective of this layer is to enable applications to recognize interesting states and to take predictive and responsive actions.

(6) **Activities**. This layer is responsible for using the contextual information from the previous level to make inferences about the states of users (for example, that the family members are asleep).

(7) **Intentions**. This layer is responsible for deciding on actions to be taken based on current activities and preferences of users.

This book focuses on the first three layers of the location stack. In this chapter, we discuss various incarnations of these layers used in existing positioning systems. In Part II of the book, we examine a particular type of sensing technology that relies on wireless signals obtained from wireless local area networks (WLANs) and discuss algorithms used to convert data from these sensors into position measurements (Chapter 6). We also discuss the benefits of data fusion in this positioning application and outline various fusion techniques for reducing uncertainty in positioning computation (Chapter 7).

3.2   Proximity detection

The first positioning technique discussed in this chapter is proximity-based positioning. As shown in Figure 3.2, these techniques determine the position of an object based on its closeness to a reference point in physical space. The reference points in this case are generally wireless transmitters, such as cellular base-stations or radio-frequency identification (RFID) transmitters. These methods operate based on the premise that radio transmitters have a limited range. Therefore, if a receiver is able to detect a valid signal from a given transmitter, the object must be within the coverage range of that transmitter. The position of the receiver is then assumed to be either the position of the transmitter, or an average of the positions of multiple transmitters within range.

A variant of proximity-based techniques considers not only the presence of a signal, but also its *strength*. The main idea behind this method is that wireless signal power decreases with distance from the transmitter (see Chapter 5 for further details on this issue). Signal-strength-based methods, therefore, assume that the distance in the received

Receiver

Base-station

Coverage area

Figure 3.2.   Proximity-based positioning.

signal space is proportional to the distance between the transmitter and the receiver in the physical space. As such, the position of the transmitter with the strongest signal (and, hence, closest to the receiver) is assigned to the receiver.

3.3 Lateration

Lateration-based methods use range (distance) measurements from multiple reference points to compute the position of a receiver. In many cases, range measurements are computed based on the knowledge of the time it takes for the signal to travel from the transmitter to the receiver (time of arrival). In particular, the distance between the transmitter and receiver is the time of travel multiplied by the speed of light. The reference points in the lateration technique are generally transmitters such as base-stations in cellular systems.

Lateration-based methods are further subdivided into circular and hyperbolic techniques, as discussed below.

3.3.1 Circular lateration

These techniques use range measurements from three or more reference points for position computation. Mathematically, circular lateration proceeds as follows. Denote the position of the receiver and the ith transmitter in two-dimensional Cartesian coordinates as (x, y) and (x_i, y_i), respectively. Further, denote the computed distance between the receiver and the transmitter as r_i. Using the Pythagoras theorem, we have

$$r_i^2 = (x - x_i)^2 + (y - y_i)^2, \qquad (3.1)$$

where $i = 1, \ldots, n$ and n is the number of transmitters (base-stations) for which range measurements are available. As shown in Figure 3.3, Equation (3.1) defines a circle of radius r_i around each transmitter where the receiver may reside. The intersection of the circles (or the solution to the above system of equations) is the position of the receiver.

A key challenge in practical systems is that range measurement may often be associated with errors. In particular, errors in synchronization between the clocks on the transmitter and receiver can lead to errors in computing the travel time and hence the range measurement. The existence of these errors means that the system of equations defined in Equation (3.1) may not always have a unique solution (that is, the circles in Figure 3.3 may not intersect at a unique point). In such situations, the least squares solution discussed below, or more advanced techniques described in [44, 50] can be used to obtain an approximate solution to the system of equations.

To compute the solution to the above system of equations using the least squares method, we note that

$$r_i^2 - r_1^2 = (x - x_i)^2 + (y - y_i)^2 - (x - x_1)^2 - (y - y_1)^2 \qquad (3.2)$$

$$= x_i^2 + y_i^2 - x_1^2 - y_1^2 - 2x(x_i - x_1) - 2y(x_i - y_1). \qquad (3.3)$$

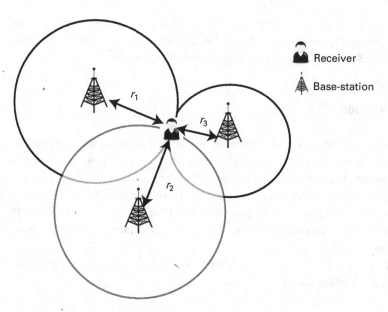

Figure 3.3. Positioning based on circular lateration.

Considering $i = 1, \ldots, n$, we can rewrite the system of equation in matrix form as follows:

$$HX = B, \tag{3.4}$$

where (with superscript T denoting the transpose)

$$X = [x \, y]^T, \tag{3.5}$$

$$H = \begin{bmatrix} x_2 - x_1 & y_2 - y_1 \\ \vdots & \vdots \\ x_n - x_1 & y_n - y_1 \end{bmatrix}, \tag{3.6}$$

and

$$B = \frac{1}{2} \begin{bmatrix} (r_1^2 - r_2^2) + (x_2^2 + y_2^2) - (x_1^2 + y_1^2) \\ \vdots \\ (r_1^2 - r_n^2) + (x_n^2 + y_n^2) - (x_1^2 + y_1^2) \end{bmatrix}. \tag{3.7}$$

Then, the least squares solutions to the system of equations can be obtained as follows [44]:

$$X = (H^T H)^{-1} H^T B. \tag{3.8}$$

Lateration in three-dimensional space can be carried out in a similar manner.

3.3.2   Hyperbolic lateration

Hyperbolic lateration methods use range differences between two references to compute the position of a receiver. Similar to the case of circular lateration, a system of equations can be composed of the range differences between every pair of reference points $\forall i, j; i \neq j$:

$$d_{ij} = r_i - r_j \tag{3.9}$$

$$= \sqrt{(x - x_i)^2 + (y - y_i)^2} - \sqrt{(x - x_i)^2 + (y - y_i)^2}, \tag{3.10}$$

where d_{ij} denotes the range difference between reference points i and j. The solution to the above system of equations can be obtained as in the case of circular lateration [50] by noting that

$$(r_1 + d_{i1})^2 = r_i^2. \tag{3.11}$$

Therefore, we have

$$(r_1 + d_{i1})^2 = r_i^2, \tag{3.12}$$

and

$$x_i^2 + y_i^2 - x_1^2 - y_1^2 - 2x(x_i - x_1) - 2y(x_i - y_1) - d_{i1} - 2d_{i1}r_1 = 0. \tag{3.13}$$

The above system of equations can be written in matrix form as follows:

$$HX = B, \tag{3.14}$$

where

$$X = [x \ y \ r_1]^T, \tag{3.15}$$

$$H = \begin{bmatrix} x_2 - x_1 & y_2 - y_1 & r_{21} \\ \vdots & \vdots & \vdots \\ x_n - x_1 & y_n - y_1 & r_{n1} \end{bmatrix}, \tag{3.16}$$

and

$$B = \frac{1}{2} \begin{bmatrix} (x_2^2 + y_2^2) - (x_1^2 + y_1^2) - d_{21}^2 \\ \vdots \\ (x_n^2 + y_n^2) - (x_1^2 + y_1^2) - d_{n1}^2 \end{bmatrix}. \tag{3.17}$$

The least squares solutions to the system of equations can be obtained again as follows [44]:

$$X = (H^T H)^{-1} H^T B. \tag{3.18}$$

As shown in Figure 3.4, the set of all points for which the range difference between the two references is constant defines a *hyperbola*. The receiver lies on the intersection of these hyperbole.

Range differences can be computed using the difference in the arrival time of signals from two transmitters to the receivers (time difference of arrival).

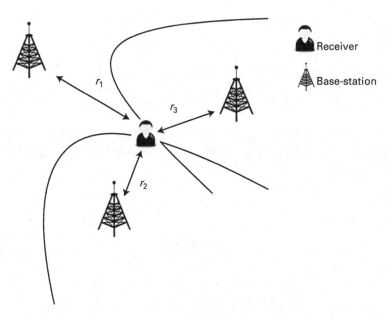

Figure 3.4. Hyperbolic-lateration-based positioning. The two hyperbole are defined by the pairs of range measurements (r_1, r_2) and (r_1, r_3). The base-stations define the foci of the hyperbole.

3.4   Angulation

Angulation-based positioning techniques employ the angle of arrival of a wireless signal to determine the position of a receiver (angle of arrival). As shown in Figure 3.5, in this technique the line connecting two reference points may be used as an internal reference. The angle between the receiver and transmitter, θ_i, can then be given by:

$$\tan \theta_i = \frac{y - y_i}{x - x_i}. \tag{3.19}$$

Therefore,

$$(x - x_i)\sin \theta_i = (y - y_i)\cos \theta_i. \tag{3.20}$$

The above defines a system of equations which can be written in matrix form as follows:

$$HX = B, \tag{3.21}$$

where

$$X = [x \ y]^T, \tag{3.22}$$

$$H = \begin{bmatrix} -\sin(\theta_1) & \cos(\theta_1) \\ \vdots & \vdots \\ -\sin(\theta_n) & \cos(\theta_n) \end{bmatrix}, \tag{3.23}$$

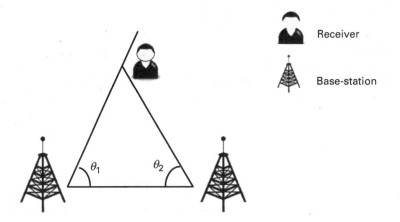

Figure 3.5. Angulation-based positioning.

and

$$B = \frac{1}{2} \begin{bmatrix} y_1 \cos(\theta_1) - x_1 \sin(\theta_1) \\ \vdots \\ y_1 \cos(\theta_n) - x_1 \sin(\theta_n) \end{bmatrix}. \tag{3.24}$$

The least squares solutions to the system of equations is given by [44]

$$X = (H^T H)^{-1} H^T B. \tag{3.25}$$

The angle of arrival can be measured in various ways, including by employing directional antennae, or through the use of the phase properties of the signal [24].

3.5 Fingerprinting

The fingerprinting approach to positioning relies on training signals from a set of reference points (anchor points) with known locations (Figure 3.6). The training data are generally collected during a *training phase*. During the on-line operation of the system, the observed signal is compared against the stored training records for each of the reference points. The position of the reference point whose training record most closely matches that of the observation is returned as the position of the receiver. Fingerprinting techniques are widely used in positioning in wireless local area networks and will be discussed in great detail in Chapter 6.

Fingerprinting-based techniques are advantageous in that they do not require any assumption regarding the nature of the propagation environment. Instead, the radio environment is modeled based on the training data. As such, these techniques are relatively more resilient to non-line-of-sight propagation than the positioning techniques discussed previously. The main disadvantage of fingerprinting techniques lies in the fact that the accuracy of fingerprinting techniques depends on the availability and accuracy of the

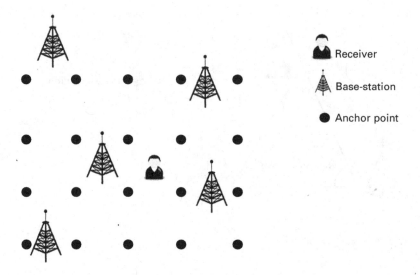

Figure 3.6.   Fingerprinting-based positioning.

training data. Training data must be collected over multiple time intervals to ensure that the time-varying nature of the propagation environment is captured.

3.6   Dead reckoning

Dead reckoning systems aim to predict the position of a user based on the user's movement history. These techniques employ measurements of motion parameters such as velocity (speed and direction), acceleration, and elapsed time as well as measurements of environmental qualities including frictional forces [24].

The main limitation of dead reckoning systems is their susceptibility to drift. Because position estimates are computed based on previous estimates, any estimation error will propagate through to future computations. As such, external information is needed to reset the estimate at regular intervals. As we shall see in Chapter 7, dead reckoning estimates are extremely valuable for complementing other position estimates.

3.7   Computer vision

Computer vision-based methods track the position of mobile users over time by employing visual representations of the environment [37, 67, 86]. The methods use one or more cameras to monitor an environment visually. Figure 3.7 shows the main components involved in vision-based tracking. Each camera captures a sequence of images of the monitored environment. These images are used to extract various visual features within the image frame to detect and identify objects of interest within the images. Finally, the

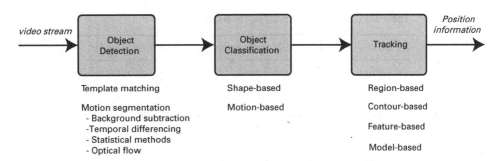

Figure 3.7. Components of a computer vision-based positioning algorithm.

position of the object of interest within the frame is tracked and translated into real-world coordinates from camera coordinates. For a detailed survey of visual tracking methods, the reader is referred to [37].

While traditional visual monitoring methods often require a human operator, recent efforts have focused on the development of tools for automatic tracking of people in video sequences. In fact, video-based tracking has been an active research area in surveillance, security, authentication, and context awareness [32, 37].

Computer vision-based techniques can provide positioning accuracies in the range of centimeters in indoor environments. Performance of these systems, however, is highly influenced by environmental conditions such as lighting and the presence of occluding objects.

3.8 Comparison of positioning techniques

Table 3.1 summarizes the measurement modalities and the corresponding sources of error. As seen, the accuracy of proximity-based methods suffers from coarse positioning granularity, which results from the quantization of the space into cells of coverage for each transmitter. Wireless positioning systems that employ laterization methods compute range measurements based on the time of arrival or time difference of arrival of radio signals. These systems are generally susceptible to ranging errors that result from non-line-of-sight propagation and synchronization issues. Moreover, these methods require specialized hardware (clocks) to compute range measurements.

Angulation methods use the direction of incoming radio signals to compute the location of a receiver. In this case, additional hardware is needed to measure the direction of the incoming beam.

Fingerprinting-based techniques rely on training data collected at a set of anchor points with known locations to model the characteristics of the radio propagation environments. Although these methods are not susceptible to non-line-of-sight and synchronization errors, their accuracy is heavily dependent on the accuracy and reliability of the training data collected at the anchor points.

Table 3.1. Comparison of positioning techniques

Method	Measurement type	Sources of error
Proximity	signal power	spatial quantization
Circular lateration	time of arrival	NLOS, synchronization errors
Hyperbolic lateration	time difference of arrival	NLOS
Angulation	angle of arrival	NLOS
Fingerprinting	received signal strength	environmental changes
Dead reckoning	acceleration, velocity	bias
Visual surveillance	video	occlusion, lighting

NLOS: non-line-of-sight

Dead reckoning systems do not use properties of wireless signals. Instead, they rely on an initial position estimate and motion parameters for positioning. The main shortcoming of these methods is positioning drift that occurs as a result of error accumulation.

Finally, computer vision-based methods employ a visual representation of the environment to determine positioning of moving objects. Though these methods can provide high positioning accuracies, their performance may be affected by changes in lighting conditions and occlusions. Moreover, computer vision-based methods generally incur a high computation cost due to the need to process and possibly store a large number of images. Scalability is also a concern in such systems since only a few users can be tracked at a time. Moreover, the cost of installation and maintenance of cameras often hinders large-scale deployment of these systems. Finally, since computer vision-based techniques process images which disclose the identity of the persons tracked, privacy and consent are important considerations in these systems.

The techniques discussed in the above sections employ different measurement modalities for positioning. As such, each technique is susceptible to different noise mechanisms. This suggests the development of systems that rely on multiple and complementary measurements to improve positioning accuracy and reliability.

3.9   Chapter summary

This chapter discussed the measurements and positioning techniques used in radio- and satellite-based positioning systems. In particular, we reviewed methods of proximity detection, lateration, angulation, and fingerprinting. With reference to the location stack model, this discussion addressed the first two layers of sensing and measurement. Proximity-based methods use signal strength measurements and a binary decision technique for converting these measurements into position estimates. In the case of lateration techniques, sensors collect range measurements based on time of arrival or time difference of arrival. The measurement layer then uses circular or hyperbolic lateration to transform these quantities into position estimates. For angulation systems, the sensors report the direction of arrival of the radio signal and the measurement layer uses triangulation for position computation. Fingerprinting methods sense signal strength values and employ a set of anchor points to compute positions.

In addition to the above methods, which rely on the properties of wireless signals for positioning, we also reviewed dead reckoning and computer vision-based techniques. The former method senses kinematic parameters such as velocity to compute position displacement over time. The latter techniques rely on visual representations of the environment (video) for tracking objects of interest.

The positioning techniques discussed in this chapter each have advantages and shortcomings in terms of positioning accuracy, coverage area, resilience to errors, and cost. The choice of positioning technique, therefore, is highly dependent on the requirements of the particular applications. Chapter 4 will discuss how these positioning techniques are implemented in several example systems.

4 Positioning systems

Chapter 3 detailed the fundamental techniques used in positioning. In this chapter, we review several examples of existing positioning systems that use these techniques. In particular, we focus on wireless positioning systems to motivate wireless local area network positioning. We discuss the Global Positioning System (Section 4.1) and cellular positioning (Section 4.2) as examples of positioning systems used in outdoor environments, and positioning based on ultrasound and infrared (Section 4.3) as examples of indoor positioning systems. We then proceed to motivate and describe positioning systems employing wireless local area networks (Section 4.4). Finally, we conclude the chapter by comparing the advantages offered by each system.

4.1 The Global Positioning System

Historical perspective

In 1957, the first artificial satellite known as Sputnik I was launched into the earth's orbit by the Soviet Union. Within days of the launch, scientists at Johns Hopkins University noted that they could determine the position of the satellite based on the Doppler shift associated with its radio transmissions. What was even more interesting was that if the position of the satellite were known, this Doppler shift could be used to determine the position of a satellite receiver on earth. This observation ushered in the development of global navigation satellite systems (GNSS).

The first operational satellite-based positioning system was Transit, also known as the Navy Navigation Satellite System (NAVSAT). This system was primarily used for military operations by the United States Navy. This system became fully operational in 1964 and remained in use until the early 1990s. Though the Transit system effectively delivered positioning accuracies within hundreds of meters, it suffered from three shortcomings [26]. First, Transit was limited to using at most five satellites simultaneously to avoid radio interference. This limitation resulted in large time windows of service unavailability (35–100 minutes) [56]. The second limitation of the Transit system was that it could only provide two-dimensional positioning estimates. Third, this system was slow, intermittent, and its accuracy was highly sensitive to the motion of the receiver [26].

Despite these limitations, the Transit system set the backdrop for the development of improved satellite-based positioning systems. In particular, in the early 1970s, the United

States Department of Defence approved the development of the Global Positioning System (GPS). Although work on the GPS began in the early 1960s, it was not until 1978 that the first experimental GPS satellite was launched [86]. Finally, in 1995 the system was declared fully operational [20]. (For a detailed history of the development of GPS, the reader is referred to [26].) The GPS was initially developed for military navigation operations. However, it was made available for civilian use after a Korean Airlines plane was shot down by a Soviet interceptor when it entered the USSR prohibited airspace due to a navigational error.

The GPS system is the best known and most commonly used global navigation system in use today. Several other systems, however, are in development. These include the Galileo system (under development by the European Union), the COMPASS system (under development by the People's Republic of China), and the GLONASS system (under development by Russia). GPS is the most widely used positioning system today, finding its way into a wide range of applications including aviation, surveying and mapping, agriculture, and navigation (both in commercial and personal contexts). GPS services are now available inexpensively for a mass market through GPS-enabled mobile phones. It is therefore not surprising that ABI Research forecasted GPS shipments to exceed 1.1 billion in 2014, with an estimated revenue of $100 billion in 2012. It is interesting to note here that the cost of development of GPS is estimated to be $10 billion and its annual operational costs are pegged at $250–$500 million [20]. The mass market availability of GPS has paved the way for the development and delivery of location-based services. In this sense, GPS has had a profound impact on the landscape of mobile computing systems.

Technical overview

The Global Positioning System consists of three segments [26] as follows.

- *Space* The space segment comprises a constellation of 24 satellites in the earth's orbit as shown in Figure 4.1 (the actual number of satellites in orbit is 31 to accommodate for spares in case of satellite failure and maintenance). The satellites are arranged in six orbital planes (at 55 degrees tilt relative to the earth's equator), such that four satellites are visible to a GPS receiver at any point on the earth at a given time. The satellites orbit the earth at an altitude of 20 200 kilometers. this allows the satellites to cover the same ground track once each sidereal day (the time for the earth to rotate 360 degrees, approximately 23 hours and 56 minutes) [50]. Each satellite transmits a signal that can be used to determine a receiver's distance from the satellite. The position of the satellites is known at all times. The signal is transmitted with a data rate of 50 bits per second on two frequencies (L1: 1.57542GHz; L2: 1.2276 GHz) [24], although additional frequency bands of L3–L5 are also used for other purposes [74].
- *User (receiver)* The receiver obtains satellite signals and is responsible for determining the user's position. As previously mentioned, GPS receiver chips may be embedded in dedicated GPS devices or mobile computing devices such as phones.

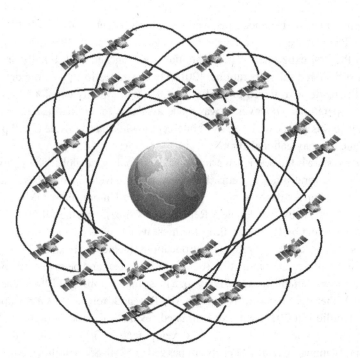

Figure 4.1.   Satellite constellation used in the GPS system.

- *Ground control stations* These stations are dispersed around the world and have accurately known locations. The stations determine and predict satellite positions and provide clock corrections.

Position estimation using GPS relies on circular lateration, which relies on range measurements. To enable this mode of positioning, the satellites transmit pilot signals carrying information needed to compute the position of satellites (for example, orbit information and clock corrections) and range measurements [50]. The latter are computed based on the transit time of the signal traveling between a GPS receiver and a satellite within line-of-site. Computation of this transit time relies on the fact that transmitted signals are modulated using known pseudo-random noise (PRN) codes (for a detailed description of these codes, the reader is referred to [50]). In particular, both the satellite and the receiver simultaneously start generating the same random code. As shown in Figure 4.2, the receiver then compares its locally generated copy of the code to that received from the satellite. The time delay between the two codes represents the travel time of the signal. This time is then multiplied by the speed of light to compute the distance (range) between a satellite and the receiver. Given four such distance measurements, circular lateration is used to compute the position of the receiver.

Given the range measurement and positions of the satellites, lateration proceeds as follows. Denote the position of the GPS receiver and the ith satellite in three-dimensional Cartesian coordinates as (x, y, z) and (x_i, y_i, z_i), respectively. Given r_i, the computed

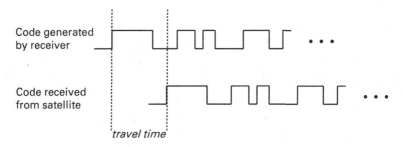

Code generated
by receiver

Code received
from satellite

travel time

Figure 4.2.   Computation of the signal travel time in the GPS system.

distance between the receiver and the satellite, we have

$$r_i = \sqrt{(x - x_i)^2 + (y - y_i)^2 + (z - z_i)^2} - c\delta t, \tag{4.1}$$

where c is the speed of light and δt is the receiver clock bias (difference between the receiver clock and the GPS time). The clock bias generally results from synchronization errors on the receiver. This is due to the use of inexpensive oscillators on the receiver (as compared to the highly accurate atomic clocks on the satellites) [56]. As such, the clock bias is the same for all satellites.

The position of the receiver can then be computed by solving a system of equations given distances to four satellites ($i = 1, \ldots, 4$) as discussed in Section 3.3.

Given ideal, line-of-sight conditions, the GPS can position a receiver with an accuracy of a few meters. Since line-of-sight communication to GPS satellites is generally not possible in an indoor environment, this positioning system is geared towards outdoor positioning. A detailed account of GPS operation is beyond the scope of this book. The interested reader is referred to the wealth of information available on this topic (see, for example, [20, 50]).

4.2   Cellular-based positioning systems

Historical perspective

In 1996, the US Federal Communications Commission introduced the Enhanced 911 (E-911) mandate, requiring cellular providers to track the location of their subscribers with prescribed accuracies. In particular, cellular providers were required to locate the mobile device to within 50 meters over 67% of the time. This positioning information is used to locate the subscriber when an emergency call is placed. Since cellular infrastructures have been equipped with positioning capabilities, the positioning information has been used to deliver value-added location-based commercial services, including location-sensitive billing and advertising, traffic control, and wireless resource allocation [63].

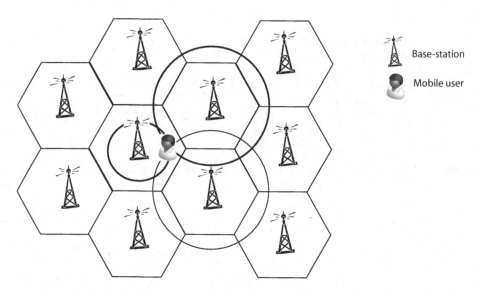

Figure 4.3.   Overview of trilateration methods in cellular-based positioning.

Technical overview

In a cellular network, the overall coverage area is divided into non-overlapping *cells* (shown in Figure 4.3). Each cell is served by a centralized *base-station*, which enables wireless telephone connection through its connection to a mobile telephone switching office [27]. Mobile phones continuously sense the signal strength from their nearest base-stations to associate with the strongest base-station. The simplest positioning method in cellular networks takes advantage of the continuous availability of signal strength measurements to locate the user. This proximity-based method is known as *cell-of-origin*. In this case, the base-station with the strongest signal is assumed to be the one closest to the mobile device spatially, and the position of the nearest base-station is used to provide coarse positioning information about the mobile. The positioning accuracy of cell-of-origin methods is clearly dependent on the size of the cells and the distance between adjacent base-stations, which ranges from a few hundred meters in urban regions to several kilometers in rural areas.

Finer-grade positioning in cellular systems can be accomplished by measuring various properties of the wireless signal, giving rise to four classes of positioning algorithms [63]. The first technique employs angle-of-arrival-based positioning. In this method, base-stations use specialized antennae to measure the bearing of signals received from mobile devices. Given known bearings to two or more base-stations, triangulation is used to locate the mobile device (as discussed in Section 3.4). The remaining three techniques compute the distance between the base-station and the mobile device based on measurements of the signal strength received at the mobile or by measuring the propagation time of the signal between the mobile and the base-station. The latter methods measure either the time of arrival of the radio beam at the base-station, or the difference between

propagation between the mobile and a pair of base-stations is measured (time difference of arrival). These measurements require the computation of the distance between the mobile and at least three base-stations. Once these distances are available, circular and hyperbolic trilateration methods are used to determine the position of the mobile device, as discussed in Section 3.3.

4.3   Ultrasound and infrared systems

Historical perspective

Positioning based on GPS or cellular techniques can provide high accuracy in outdoor environments. The utility of these systems is, however, limited in indoor environments where line-of-sight communication with satellites and cellular base-stations may not be possible. In addition, in indoor positioning applications, such as navigation in buildings, the accuracy of GPS and cellular systems is often insufficient. These factors necessitated the development of positioning systems that provide fine-grade positioning in indoor environments. Early indoor positioning systems employed light and sound waves, instead of radio waves.

Technical overview

The Active Badge system [87] developed by Olivetti Research Laboratory was one of the first positioning systems for indoor environments. In this system, each user wears a badge which periodically emits a unique infrared identifier. This identifier is sensed by infrared receivers installed in the environment. The position of the user is determined based on their proximity to known infrared receivers, and, as such, the accuracy of this system is dependent on the density of the receivers. The limitation of this system is the requirement for a specialized infrastructure comprising the infrared receivers and badges. Moreover, the infrared sensing range is limited to several meters, limiting the scalability of the system in large environments. Another limitation of the system is the possible interference between infrared signals and fluorescent light or sunlight.

A second example of a badge-based system is the Active Bat system developed by AT&T. This system relies on ultrasonic beams for positioning. It employs the time of flight of the ultrasound system together with lateration to determine the position of a user with accuracy in the order of centimeters [35]. Another example of a positioning system employing ultrasonic beams is Cricket [70], which was developed as part of MIT's Project Oxygen. In this system, beacon stations mounted on walls and ceilings broadcast a combination of radio frequency and ultrasonic waves which are received by a set of passive receivers. The receivers measure the difference in arrival time between the radio frequency and ultrasonic signals to determine the travel time of the latter (sound waves travel much slower than radio frequency waves). This distance is used to determine the closest beacon to the receiver. The Cricket system can achieve accuracies in the range of several centimeters to meters, depending on the beacon density. One of the main limitations of this system is its sensitivity to non-line-of-sight propagation [56].

Moreover, similar to infrared systems, the systems using ultrasonic beams require a specialized infrastructure for positioning.

4.4   Wireless local area network (WLAN) positioning

Historical perspective

Over the past two decades, wireless local area networks (WLANs) have become ubiquitously available in commercial and home settings. For example, many companies, airports, hotels, coffee shops, and other public spaces offer WLAN access. Moreover, most commonly used computing devices, such as personal computers, laptops, and smart phones, now have WLAN capabilities through integrated WLAN network interface cards (NICs).

These widely deployed WLAN infrastructures can be used for implementation of cost-effective positioning systems. Early work in WLAN positioning employed "cell-of-origin" methods to determine the location of a mobile device based on the nearest access point (i.e. access point with the strongest received signal strength). The first fine-grained WLAN positioning systems were introduced in 2000 (see, for example, the pioneering RADAR system [4]). Soon thereafter, WLAN positioning solutions were commercialized for use in applications such as asset tracking, resource management, and network security [73].

Technical overview

WLAN positioning is the process of determining the physical coordinates of mobile network devices, such as laptops or personal digital assistants, based on observing radio signals exchanged between the device and (WLAN) access points. In a typical WLAN positioning setup, a user operates a wireless device equipped with WLAN communication capabilities. This device exchanges probe signals with WLAN access points in the environment and measures features of radio signals relevant for positioning as detailed in Chapter 5. The most commonly used technique in these systems is fingerprinting. Part II of the book is dedicated to an in-depth discussion of this topic.

WLAN-based positioning systems can provide accuracies in the order of meters in indoor spaces. Owing to the ubiquity of WLANs, these systems can provide large coverage in both indoor and outdoor spaces with a relatively low cost.

4.5   Comparison of positioning systems

4.5.1   Evaluation criteria

The performance of a positioning system can be evaluated based on many factors. These factors include the following.

- **Accuracy** This performance measure is defined as the *closeness* of the estimated and actual positions. Clearly, the closer the estimate and actual positions, the better the performance of the system. The notion of closeness can be defined in many ways. A common measure of accuracy is the error between the estimated and actual positions in Cartesian space. Denote the estimate and actual positions in the two-dimensional space as $\hat{\mathbf{p}} = [\hat{p}_x\ \hat{p}_y]$ and $\mathbf{p} = [p_x\ p_y]$, respectively. Then, the error vector is defined as follows:

$$\mathbf{e} \triangleq \mathbf{p} - \hat{\mathbf{p}}. \tag{4.2}$$

The l_2 norm of this error vector is often used as a measure of positioning error:

$$e = ||\mathbf{p} - \hat{\mathbf{p}}|| \tag{4.3}$$

$$= \sqrt{(\hat{p}_x - p_x)^2 + (\hat{p}_y - p_y)^2}. \tag{4.4}$$

The above error is generally averaged over multiple instances of positioning to obtain an overall measure of system accuracy:

$$\bar{e} = \frac{1}{N} \sum_{i=1}^{N} e(i), \tag{4.5}$$

where $e(i)$ is the positioning error for the ith positioning estimate and N is the total number of estimates. The above measure is also known as the *bias* [65] of the position estimator, indicating the average deviation from the true position.

- **Consistency** In addition to the requirement that a positioning system provides accurate estimates, it is also important that the system provides such accurate estimates *most of the time*. This ability is termed consistency in this book. One way to measure the consistency of the system is to report the *variance* of positioning error over multiple estimates:

$$var(e) = \frac{1}{N-1} \sum_{i=1}^{N} (e(i) - \bar{e})^2. \tag{4.6}$$

- **Cost** Deployment and maintenance costs are important considerations in positioning systems. Modern positioning systems often employ mobile devices with limited power and computational resources. Therefore, it is also essential that computational demands of positioning techniques are considered in addition to deployment and operating costs.
- **Coverage** Positioning systems operate on different geographic scales (global, city level, building level, room level, etc.). Therefore, when evaluating a positioning system for a specific application, it is important that the coverage area of the system is considered.
- **Scalability** Recent years have witnessed mass market penetration of mobile devices. In the context of location-based services, scalability of positioning systems must be considered to cater to the growing number of mobile devices.
- **Privacy** As discussed in Chapter 1, privacy and consent are essential features of a positioning system.

Table 4.1. Positioning systems compared

System	Accuracy	Cost	Coverage	Privacy
GPS	m–km	high	global (outdoor)	low
Cellular	m–km	high	global (outdoor/indoor)	low
Ultrasound	cm–m	medium	building-level (indoor)	medium
Infrared	cm	medium	building-level (indoor)	medium
WLAN positioning	m	low	building-level (outdoor/indoor)	high

cm: centimeters; m: meters, km: kilometers

4.5.2   Evaluation

Table 4.1 summarizes the features of the positioning systems discussed in this chapter with respect to the above criteria.

GPS-based and cellular-based methods both rely on infrastructures that incur a high deployment and operation cost. For example, the GPS system reportedly cost $14 billion to deploy, and $500 million per year to maintain and operate [56]. While these systems can offer reliable and accurate positioning services on a global scale, privacy of location information in these systems can be compromised if service providers have access to the information.

Ultrasound and infrared methods can provide accurate positioning information in indoor environments. The main disadvantage of these systems in terms of cost is the need to deploy and maintain a dedicated infrastructure for positioning (consisting of both receivers and transmitters). This limits the scalability of the systems with the number of users. In these systems positioning may be performed locally on the receiver, allowing for increased privacy of location information. However, the need to wear badges (often visible) lowers the level of privacy offered by these systems.

With respect to the above criteria, WLAN positioning systems offer three advantages [46, 52].

- **Cost effectiveness** As part of the wireless connectivity protocol, network interface cards (NICs) measure signal strength values from all wireless access points in range of the receiver. Therefore, signals needed for positioning can be obtained directly from NICs available on most hand-held computing devices. This allows for the implementation of positioning algorithms on top of existing WLAN infrastructures without the need for any additional hardware. Due to the ubiquity of WLANs, this mode of positioning provides a particularly cost-effective solution for offering LBS in commercial and residential indoor environments.

- **Scalability** WLAN positioning systems offer scalability in two respects: (1) the cost of required infrastructure and hardware, and (2) the number of mobile devices subscribing to positioning services. Hardware scalability is due to the wide deployment of WLAN access points in commercial and residential environments as well as the increasing availability of IEEE 802.11 capabilities on personal devices such as phones and iPods. Scalability in the number of users is achieved in terminal-based positioning solutions since each mobile device is responsible for performing its own sensing and computations (although this can be achieved with satellite and cellular systems).

- **Consent and privacy** In contrast to visual surveillance, all sensing operations in WLAN positioning require the cooperation of the mobile device. Moreover, in terminal-based positioning, users must initiate positioning operations (for example, by starting a software program on the device). They can also choose to terminate positioning services by shutting off wireless communications with the infrastructure. Lastly, since positioning operations can be fully implemented on mobile clients, no invasive sensing, processing, and storage are required in WLAN positioning. Consequently, the need for implementation of additional security measures to protect position estimates is eliminated as no information is communicated over wireless links.

Due to the above advantages, WLAN positioning systems can be implemented over large areas to serve many mobile devices simultaneously in a scalable and cost-effective manner.

4.6 Chapter summary

In this chapter, we briefly discussed the development and technical aspects of several prominent positioning systems that rely on wireless signals. These system included global navigation satellite systems (the Global Positioning System in particular), cellular-based positioning systems, infrared and ultrasound systems, and finally positioning systems based on wireless local area networks (WLANs). Finally, we compared the features of these systems and suggested that WLAN-based systems can provide advantages with respect to cost, scalability, and privacy. These factors motivate further investigation of these systems. In this light, the rest of this book focuses on a detailed description of computational techniques used for WLAN positioning.

Part II

Signal processing theory

5 Positioning in wireless local area networks

In the first part of this book, we motivated the need for accurate and cost-effective positioning systems that can enable the delivery of location-based services (LBS). Moreover, we suggested positioning based on wireless local area networks as a cost-effective and reliable solution for supporting such services in indoor environments. The second part of this book provides an in-depth treatment of the technical fundamentals of these systems. This chapter will begin by providing an overview of wireless local area networks (Section 5.1) and the radio features that can be used for positioning in these networks (Section 5.2). After motivating the use of received signal strength as the feature of choice for positioning, we proceed to describe the details of the experimental data set used for illustrating various concepts in this book (5.3). This is followed by a discussion of the spatial and temporal properties of received signal strength (Section 5.4), challenges in using this radio feature (Section 5.6), and techniques for modeling the relationship between received signal strength and position (Section 5.5).

5.1 Wireless local area networks

Wireless local area networks (WLANs) are radio-based communication network infrastructures based on the IEEE 802.11 family of standards. These networks operate in the unlicensed frequency bands [27], primarily employing the 2.4 GHz ISM band. As shown in Figure 5.1, WLANs use an architecture that relies on a set of base-stations for facilitating communication among the devices within the network, and between the network and the outside world (for example, the Internet). These base-stations are known as *access points*. The coverage area of each access point is generally a few hundred meters and is known as the *Basic Service Area* (BSA). While WLANs also allow peer-to-peer (ad hoc) connections, this architecture is rarely used. All access points and mobile clients within a WLAN are uniquely identifiable through a Media Access Control (MAC) address.

The IEEE 802.11 standard allows for a variety of data rates, modulation schemes, bandwidths, and multiple access methods. The focus of this book is mainly the IEEE 802.11b and g families. The former was finalized in 1999. It uses Carrier Sense Multiple Access with Collision Avoidance (CSMA/CA), and provides data rates up to 11 Mbits/s. The IEEE 802.11g was finalized in 2003 and uses orthogonal frequency-division multiplexing (OFDM) to offer data rates of up to 54 Mbit/s.

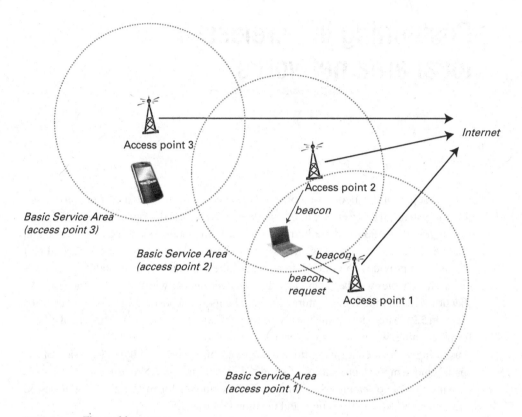

Figure 5.1.   The problem setup.

Positioning in WLAN systems relies on a set of *beacons* exchanged between the access points and the mobile clients. In particular, access points periodically broadcast beacon signals which contain various communication-related information including a time stamp, path loss information, and supported data rates. The strength of these beacon signals is measured and used in positioning. Beacon signals can be exchanged in two modes. In the *passive scanning* mode, mobile devices listen to beacon transmissions from all access points. This is done as part of the communication functions in the network to decide which access point should be used by the mobile device for communication (the access point with the strongest signal to noise ratio is chosen). The mobile device may operate in an alternative mode known as *active scanning*, where it sends probe requests to the nearby access points. Access points then respond to this probe request by transmitting beacon signals, as discussed previously.

WLAN positioning systems operate based on the knowledge of the relationship between physical positions and some feature of the beacon received at the mobile device. As we will see in later chapters, to design a positioning system, characteristics of the radio features must be known. In this chapter, we motivate the use of received signal strength (RSS) as the feature of choice in WLAN positioning. Moreover, we investigate the spatial and temporal properties of RSS.

5.2   Radio signal features in WLANs

WLAN positioning exploits the dependency between the location of a mobile device and the characteristics of beacon signals exchanged between the device and a set of physically distributed WLAN access points. As discussed in Chapter 3, four signal features may be used for positioning.

- **Angle of arrival** (AoA) This feature measures the direction of the radio wave incident on the receiver's antenna. Given two or more such angles, angulation can be used to determine the position of a mobile device.
- **Time of arrival** (ToA) This feature measures the travel time of the radio signal from the transmitter to the receiver. Since the radio signal travels at a known speed, the ToA can be used to determine the distance between the transmitter and the receiver. Given three or more such distances, circular lateration can be used to find the position of a mobile device [57].
- **Time difference of arrival** (TDoA) This feature measures the difference in ToA at two different receivers. Similar to ToA positioning, three or more TDoA measurements can be used to locate a mobile device using hyperbolic lateration.
- **Received signal strength** (RSS) As the name suggests, this feature measures the radio signal power received at the mobile device.

Although AoA has been used in WLAN positioning, its measurement requires specialized antennae leading to additional hardware cost. Moreover, ToA and TDoA require precise synchronization between transmitter and receivers, which is very difficult to achieve in WLANs [50]. As discussed previously, the RSS is measured by the mobile device to select the most appropriate access point for communication. As such, this feature is easily obtainable without the need for any additional hardware. This makes RSS the feature of choice in most WLAN positioning systems.

Received signal strength decreases with increasing distance between the transmitter and receiver. Therefore, the location-dependency of RSS can be used to locate a mobile device effectively. This, however, requires accurate modeling of the properties of RSS. The ensuing sections discuss these properties in detail. Before proceeding to this discussion, we introduce an experimental setup used to obtain example data for the illustration of concepts in this book.

5.3   Characteristics of the example environment

Throughout this book, we will use example RSS measurements from a real environment to illustrate various concepts. In this section we describe the method used for collecting the sample data sets. It is important to note that these data sets are collected in a real environment and not an artificially constructed setup. Using a real communication infrastructure enables us to identify practical limitations and difficulties that are encountered in WLAN positioning systems.

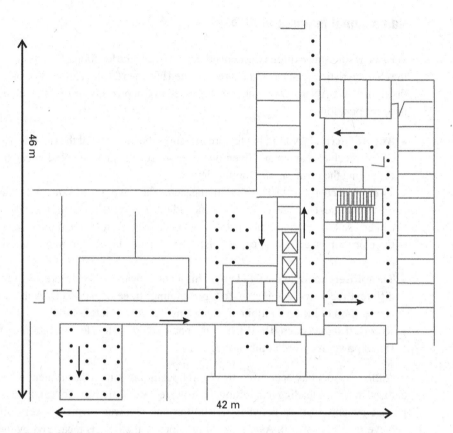

Figure 5.2.   Map of the experimental area.

Figure 5.2 shows the map of the experimental environment. The measurements used herein are collected on the fourth floor of an eight-story building which contains rooms and hallways of various sizes. In this book, we will use two separate data sets to reflect two different usage scenarios for WLAN positioning. The first data set is collected while the mobile user remains stationary at each test point. This scenario corresponds to the use of portable, but stationary, mobile devices such as laptops. A detailed description of this data set is found in [52]. For the second data set, a user carrying a mobile device roams the environment following various motion paths. A detailed description of this set is provided in [54].

RSS measurements were collected using a Toshiba Satellite laptop with a Pentium M processor, an onboard Intel PRO/Wireless 2915ABG Network Adapter, and Windows XP operating system. RSS readings were obtained at the rate of 2 samples/s by a publicly available network sniffer software, NetStumbler.[1] RSS measurements are reported as integers in the range $(-100, 0)$ in units of decibels relative to 1 milliwatt (dBm).

---

[1] http://www.netstumbler.com

5.4   Properties of received signal strength

In an ideal propagation medium (i.e. free-space), signal power falls inversely proportional
to the square of the distance between the transmitter and the receiver. Therefore, given
measurements of transmitted and received powers, the distance between the transmitter
and the mobile device can be determined. In real environments, however, the wireless
radio channel is affected by noise, interference, and other channel impediments. Therefore, in addition to the distance traveled, two mechanisms contribute to variations in the
propagation channel and therefore affect the received signal strength. These mechanisms
are large-scale and small-scale propagation effects [84]. These phenomena cause unpredictable variations in RSS and thus complicate the task of positioning. In this chapter,
we consider the spatial and temporal properties of RSS in real environments.

5.4.1   Spatial properties

As previously mentioned, in free-space, the power of a signal radiated by a transmitter
decreases with the square of the distance between the transmitter and the receiver. In
real environments, however, this decrease is affected by large-scale channel propagation
effects, which include path loss and shadowing effects. Path loss relates to dissipation
of signal power over distances of 100–1000 meters. Shadowing results from reflection,
scattering, and absorption caused by obstacles between the transmitter and the receiver,
and occurs over distances proportional to the size of the objects in the environment [27].
Because the nature and location of the blocking objects are uncertain and may change,
the effects of shadowing are often modeled statistically. One way to do this is to assume a
log-normal distribution for the ratio of transmit-to-receive power. The combined effects
of the path loss and shadowing can then be expressed by a simple model as follows [27]:

$$P_r(\text{dB}) = P_t(\text{dB}) + 10\log_{10} K - 10\gamma \log_{10}\left(\frac{d}{d_0}\right) - \psi(\text{dB}). \tag{5.1}$$

Equation (5.1) is known as a *path loss model*. In this equation, K is a constant relating
to antenna and channel characteristics, d_0 is a reference distance for the antenna far-field,
and γ is the path loss exponent. Typical values of this parameter are $\gamma = 2$ for free-space
and $2 \leq \gamma \leq 6$ for an office building with multiple floors. Finally, $\psi \sim \mathcal{N}(0, \sigma_\psi^2)$ reflects
the effects of log-normal shadowing in the model. In indoor areas, the materials used for
walls and floors, the number of floors, the layout of rooms, the location of obstructing
objects, and the size of each room have a significant effect on path loss. As a result,
training data are often used to estimate these parameters [76, 79].

Equation (5.1) provides a simplified model for the distance–RSS relationship. In
indoor environments, however, there are several practical issues that are not reflected in
this model. First, the constant signal strength contours are generally anisotropic [49, 89].
That is, increasing distance from an access point results in RSS attenuation at different
rates in different directions. This phenomenon can be attributed to the asymmetry of
the propagation environment (walls, furniture, doors, etc.). To illustrate these issues,

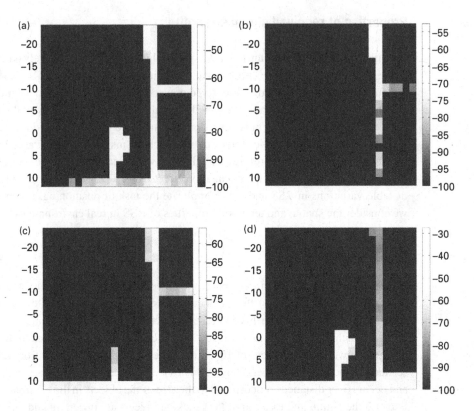

Figure 5.3.   Example of spatial RSS distributions. RSS values are averaged over 100 samples.

Figure 5.3 shows examples of the spatial distribution of mean RSS received from different access points. These examples illustrate the asymmetry of RSS distributions over space. In Chapter 6, we will discuss non-parametric techniques that use training measurements at predefined points to overcome the modeling shortcomings of the log-normal model.

Another feature of WLANs is that the number of access points providing coverage to the mobile device varies over space. To illustrate this, Figure 5.4 shows the distribution of the number of access points in the environment. In this setup, the minimum coverage occurs in a laboratory surrounded by elevators and concrete walls.

As we shall see later, most positioning systems model RSS from various access points as a vector of values, where each entry corresponds to one access point. The observation that the number of access points changes over space has important implications in these systems.

A final consideration in using the path loss model described above is that orientation of the mobile station with respect to the access point severely affects the RSS. For example, at a fixed location, a difference of nearly 9 dBm in RSS is reported for various orientations in [43]. One factor contributing to this difference is signal absorption by the

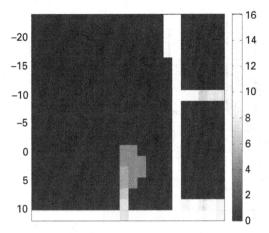

Figure 5.4. Distribution of the number of access points over the experimental area.

human body positioned between a receiver and a transmitter. The dependence of RSS on orientation further increases the complexity of the distance–RSS relationship.

5.4.2 Temporal properties

In addition to the large-scale propagation effects discussed above, small-scale fading effects also produce unpredictable variations in RSS. Small-scale fading happens when a transmitted signal is refracted and reflected by objects in the environment, causing it to reach the receiver from multiple paths. This results in multipath fading, which occurs due to constructive and destructive addition from multiple signal paths (multipath). This phenomenon happens over distances on the order of the carrier wavelength (for WLANs, $\lambda = c/2.4\,\text{GHz} = 12.5\,\text{cm}$) [27]. The envelope of the time-varying channel response in the presence of multipath fading can be modeled as a Rayleigh distribution if a strong line of sight exists between the receiver and transmitter and as a Rician distribution otherwise [27]. These effects are not explicitly considered in WLAN positioning due to modeling complexities that would result from the time-varying environment.

An important manifestation of small-scale propagation is the time variance of the channel. This time variance is due to the motion of the receiver, or the objects in the environment (for example, people), that results in changes in the location of the reflectors in the propagation path over time. It is important to note that the dependency between RSS measurements and distance varies with time because of the time-dependent changes in the environment, which lead to interference, shadowing, NLOS propagation, and multipath effects. In other words, the dynamic nature of the indoor environment leads to temporal variations in the received signal at fixed locations. Figure 5.5 depicts an example of RSS variations collected at a fixed location. As seen here, over a short time window of 200 seconds, RSS values vary as much as 20 dBm. Therefore, the received signal strength is not only dependent on the distance between the mobile user and the access

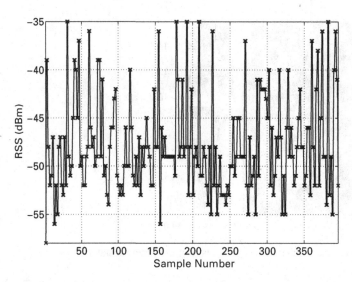

Figure 5.5.   Example of RSS measurements over time at a fixed location (sampling period = 0.5 s).

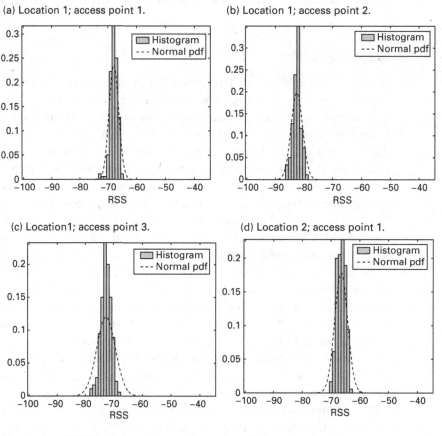

Figure 5.6.   Distribution of RSS from three access points at three different locations.

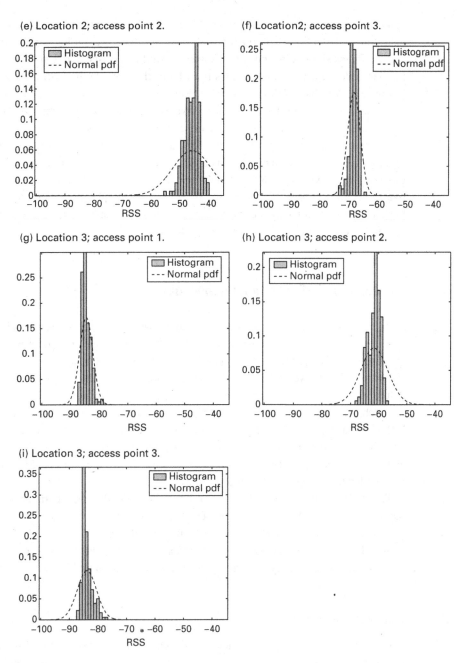

Figure 5.6. (cont.)

point, but also on time-dependent changes in the environment. This further challenges the modeling of the position–RSS dependency in real environments.

Results of measurements over various time periods presented in [43, 89] show that the temporal distributions are reported to be left-skewed (as a result of range limitations) and possibly multimodal in the presence of users. Figure 5.6 shows examples of RSS distributions from three different access points at three different locations in our setup. The examples show left- and right-skewed as well as Gaussian and multimodal distributions. Two important observations must be made here: (1) the RSS does not always follow a known distribution family (for example, Gaussian), and (2) the RSS distributions vary with location. This example motivates the need for development of data-driven techniques to model the temporal variations of RSS, as discussed in Chapter 6.

Another important property of the RSS is non-stationarity. A stationary stochastic process is one whose probability density function does not change over time. Due to the time-varying nature of the RSS environment, however, statistics of RSS tend to change over time. In fact, the RSS is known to be stationary only over short time intervals [43]. This non-stationarity means that any model of RSS distribution must be regularly updated to ensure it reflects the true properties of RSS at a given time.

5.5   Modeling the RSS–position relationship

As mentioned in previous chapters, positioning based on wireless signals relies on the fact that the relationship between some feature of the signal and the position of the mobile device is known. In the case of RSS-based WLAN positioning, this means that the relationship between RSS and receiver position must be known. Due to the complex nature of the indoor propagation environment, modeling this relationship is one of the key technical challenges in WLAN positioning. The main difficulty is that both spatial and temporal aspects of the RSS–position relationship change with time due to channel impediments, such as shadowing and multipath effects caused by moving objects. In addition, movement and orientation of the mobile device can lead to deviations from well-understood models of radio propagation.

The RSS–position dependency can be characterized using parametric or non-parametric models. Parametric approaches rely on theoretical models of radio propagation and use training data to estimate a finite number of parameters for the assumed model. In contrast, non-parametric techniques do not assume a prior model for the RSS–position dependency. Instead, they use training data to build such a model.

5.5.1   Parametric modeling

The parametric positioning approach uses the path loss model of Section 5.4.1 to model mathematically the relationship between RSS and the position of the mobile device. As previously noted, the model of Equation (5.1) does not reflect two characteristics of RSS in real environments. First, this model assumes isotropic RSS contours – an assumption often violated due to asymmetric distribution of blocking objects, such as walls and furniture. Second, this model is invariant to receiver orientation [43]. As discussed earlier in this chapter, RSS values are highly dependent on the orientation of the device [43].

An important assumption of the above model is that the locations of the access points are known exactly. This may be impractical in large and ubiquitous deployments where multiple wireless networks, run by different operators, coexist.

5.5.2 Fingerprinting-based methods

The channel impediments discussed previously result in dynamic and unpredictable propagation characteristics [38] in indoor environments. This means that, in many cases, the RSS–position relationship cannot be modeled explicitly using a functional relationship. To overcome this limitation, the second approach to modeling the RSS–position dependency is to build this model entirely from a set of training data. This training-based method is known as *location fingerprinting*, *scene matching*, or *pattern matching* [4, 10, 40, 42, 46, 52, 55, 91]. This method uses training RSS measurements that are collected at a set of *anchor points* with known positions. The collection of RSS measurements at anchor points that are distributed over the environment provides an implicit model of the spatial variations of RSS. Moreover, at each anchor point, a number of training samples are collected over time to allow the characterization of the temporal properties of RSS.

The collection of the anchor points and the RSS training data is known as a *radio map*. This map is mathematically defined as

$$\mathcal{R} \triangleq \{(\mathbf{p}_i, \mathbf{F}(\mathbf{p}_i)) \mid i = 1 \ldots N\}, \tag{5.2}$$

where $\mathbf{p}_i \triangleq [p^x \ p^y]^T$ is the Cartesian coordinates of the ith anchor point, $\mathbf{F}(\mathbf{p}_i) \triangleq [\mathbf{r}_i(1) \ldots \mathbf{r}_i(n)]$ is a fingerprint matrix [52], and n is the number of training samples collected at each anchor point. The vector $\mathbf{r}_i(t) \triangleq [r_i^1(t), \ldots, r_i^L(t)]$ contains the RSS measurements from L access points at time t at spatial point $\mathbf{p}_i$.

In constructing a radio map, two design parameters must be considered.

- **Number and placement of anchor points** As we will see later, the number of anchor points directly affects the accuracy and complexity of the positioning system. In particular, anchor points serve as known reference points for positioning (similar to satellites in GPS, transmitters in ultrasound and infrared positioning, and heavenly bodies in ancient navigation). A higher density of these points can lead to higher accuracies. However, increasing the number of anchor points will increase the labor involved in data collection and the computational complexity of the system.

    With regards to the placement of anchor points, the simplest scenario employs uniformly distributed points across the environment. Such a design, however, is often challenged by the presence of furniture and room layout that hinder signal collection at certain points in the environment. Therefore, in practice, anchor points are often distributed non-uniformly, leading to varying positioning accuracies over the physical space.

- **Number of samples collected at each anchor point** As discussed earlier in this chapter, RSS characteristics vary over time, even at a fixed spatial location. To model such time variations, multiple RSS samples are collected at each anchor point. The number of samples collected at each point depends on how fast the RSS characteristics change in a given environment. This number may also affect complexity and accuracy of the radio map, as we will see shortly.

The radio map is usually constructed off-line before the operation of the positioning system. However, as previously mentioned, indoor environments often change over time (for example, due to moving people and furniture), resulting in changes in the propagation environment. Such changes mean that observed RSS properties may often deviate from those modeled by the radio map. In fact, the sensitivity of the radio map to the changes in the environment poses an important challenge to the design of fingerprinting systems. To overcome this challenge, the radio map must be periodically updated. To reduce the labor associated with data collection, especially in large environments, RSS values can be automatically adapted. This can be achieved by deploying a set of reference RSS receivers to sense the environment continuously and adapt the entire radio map accordingly through interpolation and regression analysis [40, 49].

Another limitation of fingerprinting approaches is the dependence of RSS on the network card used for measurements. The energy reading at the antenna of the mobile client is continuous in nature and is converted to an integer value by the network card. Since RSS readings are intended to be used internally and in a relative manner for network functions, the IEEE 802.11 standard specifies neither how this mapping is done nor the accuracy of reported RSS values. Consequently, this mapping is performed based on vendor-specific conventions, leading to variations in reported RSS values from different cards. For example, the experimental observations of [41] indicate variations of up to 27 dBm between network cards made by Intel and D-Link. To mitigate the discrepancies arising from the use of different network cards, the work of [48] proposes the use of signal strength ratios between pairs of access points as opposed to RSS for fingerprinting.

Despite the above limitations, location fingerprinting provides an effective approach for modeling the RSS–position dependency for WLAN positioning. As a result, this is the method of choice in most of the existing literature [4, 10, 40, 42, 46, 52, 55, 91] and is the focus of this book.

5.6   Technical challenges in RSS-based positioning

Location fingerprinting is generally very effective in providing an implicit model of the RSS–position relationship. However, the use of this method limits the applicability of classical signal estimation algorithms that often rely on an explicit model between the measurement (RSS) and the unknown quantities (position). As such, new techniques must be used that cater directly to this implicit and training-based modeling. Chapter 6 will focus on the development of such techniques using non-parametric statistical tools.

Broadly speaking, fingerprinting methods match the incoming RSS readings from a mobile device to the collection of training RSS samples at anchor points. The positions of the anchor points that best *match* the incoming observation are used to compute the position of the mobile device. Due to uncertainties associated with RSS measurements and time variations in the environment, however, the matching process is non-trivial. In particular, these variations often lead to a many-to-many RSS–position relationship. Moreover, environmental variations, such as the presence of people and movement of furniture, may cause RSS characteristics observed during the on-line operation of the system to deviate from those learned based on location fingerprinting. As such, positioning estimation techniques relying on RSS must be resilient to these changes. In the remaining chapters of this book, we discuss various techniques for position estimation in the presence of such uncertainties. These include the use of auxiliary information (such as motion kinematics of mobile devices) and appropriate access point and anchor point selection strategies.

5.7 Chapter summary

In this chapter, we provided a general overview of wireless local area networks (WLANs) and motivated the use of received signal strength as the radio feature used for positioning. We also examined the spatial and temporal properties of received signal strength (RSS). We noted that the indoor propagation channel is complex and susceptible to various time-varying impediments. As a result, positioning systems generally model the RSS–position relationship in an implicit manner using training data collected at a set of spatially distributed anchor points. This method, known as location fingerprinting, can effectively model the complex RSS–position relationship in indoor environments, but renders existing estimation and filtering methods inapplicable. In particular, the lack of an explicit, functional relationship between RSS and position differentiates RSS-based WLAN tracking from classical target tracking problems. Chapter 6 discusses non-parametric estimation techniques that cater specifically to the implicit representation of the RSS–position relationship.

6 Memoryless positioning

The objective of a positioning system is to determine the position of a mobile device. This position, however, is not directly observable and must be determined based on some observable measurement. In the case of RSS-based positioning, this observable measurement is the received signal strength (RSS) at the mobile device. If the relationship between the RSS values and the position of mobile devices were known, the positioning problem would be trivial. However, as discussed in Chapter 5, this relationship is not deterministic in practice, but depends on the stochastic characteristics of the propagation environment. Consequently, the unknown position can only be *estimated* using RSS measurements. The focus of this chapter is the various estimation methods used to accomplish this task. In particular, this chapter focuses on *memoryless* estimators, which rely on an RSS measurement at a given time to compute the position estimate at that time. In other words, memoryless estimators do not consider the past history of user positions or RSS measurements during estimation.

We begin this chapter by developing a mathematical formulation of the memoryless positioning problem (Section 6.1) and show that this problem reduces to a density estimation problem. We next review two methods for density estimation based on the implicit training information provided in the radio map (Section 6.2). Using these density estimation techniques, we proceed to develop several position estimators (Sections 6.3 and 6.4). Finally, we conclude the chapter with the presentation of some experimental results (Section 6.5).

6.1 The problem of statistical memoryless positioning

In the fingerprinting approach, the position of a mobile device is determined by matching an RSS value observed at the device to the values stored in the radio map. In an ideal world, the observed RSS value would match the values in the radio map exactly, making the matching procedure trivial. In practice, however, the noisiness of the propagation environment and measurement equipment, as well as the time-varying nature of the RSS, make the matching very challenging. This book focuses on the use of statistical estimation techniques to address this challenge.

Figure 6.1 shows a general overview of the statistical estimation framework. Here, the objective is to estimate the *state of nature*. In our case, the state of nature is the position of a user carrying a mobile device. The challenge is that the state of nature is

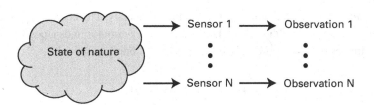

Figure 6.1.   Statistical estimation framework.

not observable. For example, we cannot observe the position of the user directly (if we could, the process of positioning would not be needed). Instead of observing the state, we can have access to a set of measurements that are correlated in some way with the state of nature. In positioning, these measurements are RSS readings from the mobile device. These measurements are related to the position of the receiver and can therefore be used to solve the positioning problem.

Statistical estimation provides a powerful tool set to model the uncertainties in the RSS values in mathematical form. In this framework, the RSS and position of the mobile device are modeled as random variables or stochastic processes, whose probability density functions are estimated using the training data stored in the radio map. Denote the RSS vector and position of the mobile at time k as $\mathbf{r}(k)$ and $\mathbf{p}(k)$. The estimated position of the mobile at this time is denoted as $\hat{\mathbf{p}}(k)$. For notational convenience, the time dependence of RSS and position is omitted in this chapter, and the symbols $\mathbf{r}$, $\mathbf{p}$, and $\hat{\mathbf{p}}$ are used instead of $\mathbf{r}(k)$, $\mathbf{p}(k)$, and $\hat{\mathbf{p}}(k)$.

Since RSS and position are dependent variables, knowledge of one can be used to predict the other. The main technical challenge is to design an estimator that takes in an RSS value and produces a position estimate that is "optimal" in some sense. In designing this estimator, the following two questions must be addressed.

- What optimality criterion should be used to judge the "goodness" of the estimate?
- How do we statistically model the RSS–position relationship using the radio map? Such a representation is needed as an explicit functional relationship between RSS and position is not available.

The rest of this section discusses the answers to these questions.

6.1.1   Optimality criteria

Two optimality criteria are generally used in positioning: (1) maximization of the likelihood density, and (2) minimization of the mean square error.

Maximization of the likelihood density
Denote the density of the RSS conditioned on the position of the mobile device as $f(\mathbf{r}|\mathbf{p})$. This density represents the probability of observing a measure $\mathbf{r}$ at position $\mathbf{p}$ and is known as the *likelihood density*. This density contains all the relevant information about the position supplied by the data [7]. The first optimality criterion considered in

this chapter is the maximization of the likelihood density. The choice of this criterion leads to a maximum likelihood (ML) estimator. This estimator finds a position estimate that maximizes the probability of the measured RSS being observed:

$$\hat{\mathbf{p}}_{ML} = \arg\max_{\mathbf{p}} f(\mathbf{r}|\mathbf{p}). \tag{6.1}$$

As we shall see later, the maximum likelihood estimator shown above chooses one of the anchor points as the estimated position of the mobile. Using this approach, however, positioning accuracy is limited by the density of the anchor points. To overcome this limitation, an extension of the above maximum likelihood estimation method forms the position estimate as the average of K anchor points with the largest likelihood densities. This method is known as the *K-nearest-neighbor* estimator. Mathematically,

$$\hat{\mathbf{p}}_{KNN} = \sum_{i=1}^{K} \mathbf{p}_{(i)}, \tag{6.2}$$

where $\mathbf{p}_{(i)}, i = 1, \ldots, K$, denotes the ordering of the anchor points based on the likelihood density.

A coarse approximation to the above estimators is proximity-based WLAN positioning. This method estimates the likelihood for each anchor point as the distance between the fingerprint at this location and the RSS observation. This approach assumes that closeness in the RSS space translates to closeness in the physical space. Alternatively, a finer-grade approximation of the likelihood density $f(\mathbf{r}|\mathbf{p})$ can be computed from the radio map data. This can be achieved using non-parametric density estimation techniques as discussed in Section 6.2.

Minimization of the mean square error

An an alternative to maximizing the likelihood density, we can aim to minimize positioning error during estimation. This leads to a minimum mean square error (MMSE) estimator that minimizes the expected l_2 norm of positioning error, i.e.

$$\hat{\mathbf{p}}_{MSE} = \arg\min_{\hat{\mathbf{p}}} \mathbb{E}\{(\mathbf{p} - \hat{\mathbf{p}})^T (\mathbf{p} - \hat{\mathbf{p}})|\mathbf{r}\}, \tag{6.3}$$

where $\mathbb{E}\{\mathbf{x}\} \triangleq \int \mathbf{x} f(\mathbf{x}|\mathbf{r})\mathbf{x}$ is the conditional expectation of $\mathbf{x}$. It can be shown that the position estimate that minimizes Equation (6.3) is the posterior mean defined as follows [65]:

$$\hat{\mathbf{p}} = \mathbb{E}\{\mathbf{p}|\mathbf{r}\} = \int_{\mathcal{P}} \mathbf{p} f(\mathbf{p}|\mathbf{r})d\mathbf{p}, \tag{6.4}$$

where $\mathcal{P}$ represents the two-dimensional indoor space where positioning is carried out. The key challenge in this estimation problem is that the density $f(\mathbf{p}|\mathbf{r})$ is unknown.

Using Bayes theorem, this density can be obtained as follows:

$$f(\mathbf{p}|\mathbf{r}) = \frac{f(\mathbf{r},\mathbf{p})}{f(\mathbf{r})} \tag{6.5}$$

$$= \frac{f(\mathbf{r}|\mathbf{p})f(\mathbf{p})}{\int_{\mathcal{P}} f(\mathbf{r}|\mathbf{p})f(\mathbf{p})d\mathbf{p}}. \tag{6.6}$$

The density $f(\mathbf{r},\mathbf{p})$ is the joint density of RSS and position and $f(\mathbf{p})$ reflects the system designer's subjective belief about the position of the mobile device *prior* to receiving any RSS observations [65]. The Bayes theorem (of conditional probability) allows one to modify this belief *after* observing the measurements based on the likelihood density $f(\mathbf{r}|\mathbf{p})$.

Substituting Equation (6.6) into Equation (6.4), we obtain

$$\hat{\mathbf{p}} = \int_{\mathcal{P}} \mathbf{p} \frac{f(\mathbf{r}|\mathbf{p})f(\mathbf{p})}{f(\mathbf{r})} d\mathbf{p}. \tag{6.7}$$

We now see that the problem of MMSE estimation reduces to estimation of either the joint density or the likelihood and prior densities.

6.1.2 Statistical radio map model

As previously mentioned, maximum likelihood and minimum mean square error estimators require knowledge of the densities $f(\mathbf{r}|\mathbf{p})$ or $f(\mathbf{r},\mathbf{p})$. The methods for estimation of these densities can be grouped into two classes: parametric and non-parametric density estimation techniques.

Parametric estimation methods assume a model for the density function and aim to estimate the parameters of the model in an optimal way. For example, if the unknown density is assumed to be Gaussian, the mean vector and covariance matrix is estimated. In situations where prior modeling assumptions for the density functions are unavailable or unreliable, non-parametric methods provide density estimates that rely solely on the underlying structure of the sample data. Since parametric forms for the likelihood density are difficult to obtain or are unavailable for the WLAN positioning problem, non-parametric techniques are generally employed. These approaches estimate the likelihood density obtained from the radio map information without the assumption of any prior statistical forms.

6.2 Density estimation

Non-parametric density estimation relies on the structure underlying a set of observed values from a density (training data) rather than prior model assumptions. These methods are especially appropriate for the WLAN position problem where RSS distributions are known not to fall into standard distribution families. This is illustrated in Figure 6.2 where histograms of RSS samples collected over time are shown for two different anchor points.

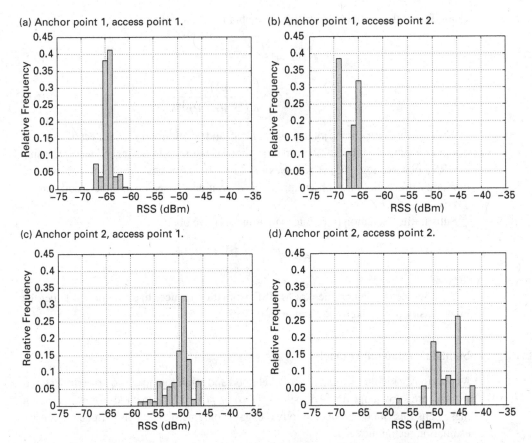

Figure 6.2.   Examples of RSS histograms for different access points at two different anchor points (320 RSS samples are used to build each histogram).

In this section we will present two non-parametric methods for density estimation. These methods are histogram and kernel density estimation.

6.2.1   The histogram

The oldest and most widely used non-parametric density estimator is the histogram [78]. Histograms have been used in widely in WLAN positioning to provide discrete approximations to the density $f(\mathbf{r}|\mathbf{p})$ [31, 55, 90, 92] by estimating the density of RSS values for each anchor point in the radio map $f(\mathbf{r}|\mathbf{p}_i)$. The ensuing sections will detail the mathematical concepts used in histogram estimation.

Univariate histogram estimate

Let us start with RSS measurements from a single access point. At anchor point i, the radio map stores $r_i(1),\ldots,r_i(n)$ measurements from this access point. Given an origin and a bin width, a histogram is constructed by counting the number of points falling into each bin.

Mathematically, denote the set of RSS measurements collected at anchor point $\mathbf{p}_i$ as $r_i(1),\ldots,r_i(n)$. Assume the data come from a sequence of identically distributed random variables. For a choice of bins $\{b_k, -\infty < k < \infty\}$, the histogram density estimate is defined as [75]:

$$\hat{f}(r|\mathbf{p}_i) = \frac{1}{nh} \sum_{t=1}^{n} I_{[b_k,b_{k+1})}(r_i(t)), \qquad (6.8)$$

where $h = b_{k+1} - b_k$ is the bin width, and

$$I_{[b_k,b_{k+1})}(x) = \begin{cases} 1 & \text{if } b_k \leq x < b_{k+1} \\ 0 & \text{otherwise.} \end{cases} \qquad (6.9)$$

The two parameters characterizing the histogram are its origin and bin width [78]. Although these parameters can have a significant effect on the density estimate, their choice has not received any attention in the context of temporal RSS characterization.

Histogram bin width and origin
In the case of a Normal density, the rule of thumb for choosing the bin width is given by [75]:

$$\hat{h} = 3.5 \hat{\sigma} n^{-1/3}, \qquad (6.10)$$

where $\hat{h}$ is the estimated bin width, $\hat{\sigma}$ is an estimate of the standard deviation of the data, and n is the number of points used to estimate the density. Since RSS densities are generally not Normal, this rule should be used with caution.

Another way to determine the histogram bin width is to consider that RSS measurements provided by network interface cards are generally quantized to integer values and the quantization step may be a natural choice for the bin width. It is, however, possible to choose large bin widths to provide insensitivity to small variations in RSS. Figure 6.3 shows the effects of histogram bin width and origin on density estimates of RSS measurements.

Multivariate histogram estimate
The histogram estimate of Equation (6.8) assumes that the RSS vector is one-dimensional. Now consider the multivariate cases where RSS values are measured from L access points at each anchor point. As such, for each access point, the radio map will store the set of measurements $\mathbf{r}_i(1),\ldots,\mathbf{r}_i(n)$. As before, $\mathbf{r}_i(t) = [r_i^1(t),\ldots,r_i^L(t)]$. The multivariate histogram estimate is constructed by assuming that RSS values received from different access points are independent:

$$\hat{f}(\mathbf{r}|\mathbf{p}_i) = \prod_{a=1}^{L} \hat{f}(r^a|\mathbf{p}_i). \qquad (6.11)$$

Cautionary notes
The histogram density estimate as formulated above suffers from several disadvantages.

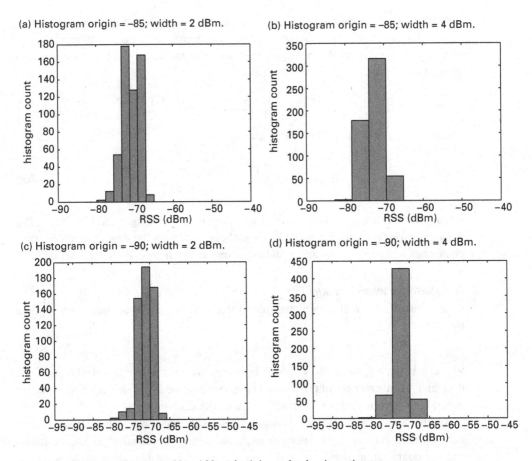

Figure 6.3.   Effects of histogram bin width and origin on the density estimate.

- The histogram suffers from the curse of dimensionality since the number of bins grows exponentially with the dimension. Furthermore, as the dimension grows, the histogram provides reasonable estimates only near the modes and away from tails of the distribution [75].
- The estimated density is discontinuous at cell boundaries and is zero outside the range of the data. Moreover, a histogram estimate at a given point only uses information localized to a particular bin; that is, observations barely falling outside that bin do not contribute to the estimate [9].
- The shape of the estimated density is highly sensitive to the choice of origin [39].

Section 6.2.2 considers an alternative approach to density estimation that overcomes some of the above shortcomings.

6.2.2   Kernel density estimator

As an alternative to the histogram, the likelihood density for each anchor point can be estimated using a kernel density estimator [75, 78]. This estimator has proved to be a promising tool in location estimation with NLOS propagation conditions in cellular networks [62], especially in the context of WLAN RSS [52, 54, 73, 89].

Kernel density estimate of the likelihood

The kernel density estimate (KDE) is the superposition of "bumps" whose shapes are determined by a kernel function [78]. Each bump is centered at a training data point. Given a set of independent and identically distributed RSS samples $\{r_i(t)|t=1\ldots n\}$ at anchor point p_i, the kernel density estimate is given by

$$\hat{f}(\mathbf{r}|\mathbf{p}_i) = \frac{1}{n\sigma_\mathbf{r}} \sum_{t=1}^{n} K\left(\frac{\mathbf{r}-\mathbf{r}_i(t)}{\sigma_\mathbf{r}}\right), \tag{6.12}$$

where $K(\cdot)$ is a non-negative, zero mean, kernel function with unit area. The parameter $\sigma_\mathbf{r}$ is known as the smoothing parameter, window width, or bandwidth [68]. This parameter serves a purpose similar to the histogram bin width. The choices of the kernel function and kernel bandwidth are important design considerations that can significantly affect positioning results. These choices are discussed next.

Kernel function

The kernel function is a d-dimensional function that is unimodal and smooth (d is the number of access points). This function must satisfy four conditions [75].

(1) The kernel function is non-negative. Mathematically:

$$K(\mathbf{r}) \geq 0, \forall \mathbf{r} \in \mathbb{R}^d. \tag{6.13}$$

(2) The area under the kernel function has unit area:

$$\int_{\mathbb{R}^d} K(\mathbf{r}) = 1. \tag{6.14}$$

(3) The kernel function has a mean of zero:

$$\int_{\mathbb{R}^d} \mathbf{r} K(\mathbf{r}) = 0. \tag{6.15}$$

(4) The function has unit variance:

$$\int_{\mathbb{R}^d} \mathbf{r}\mathbf{r}^T K(\mathbf{r}) = \mathbf{I}_d, \tag{6.16}$$

where $\mathbf{I}_d$ is the $d \times d$ identity matrix.

Three well-known examples of kernel functions include the Epanechnikov, triangular, and Gaussian kernels. The Epanechnikov kernel is defined as follows:

$$K(x) = \frac{3}{4}\left(1 - \frac{1}{5}x^2\right)/\sqrt{5} \text{ for } x < \sqrt{5}, \text{ and } 0 \text{ otherwise.} \tag{6.17}$$

The triangular kernel function is defined as:

$$K(x) = 1 - |x| \text{ for } |x| < 1, \text{ and } 0 \text{ otherwise.} \tag{6.18}$$

Finally, the Gaussian kernel is defined as:

$$K(x/\sigma_x) = (2\pi)^{-d/2} |C|^{-1/2} \exp\left(-\frac{x^2}{2\sigma_x}\right). \tag{6.19}$$

The Gaussian kernel is used in most applications, including WLAN positioning. As we shall see, the use of this function can significantly simplify mathematical computations [62, 73]. In general, the shape of the kernel function does not significantly impact the accuracy of the density estimate.

Note that the kernel functions defined above are univariate (that is, the input is a scalar). A multivariate kernel can also be constructed as the product of univariate kernels. In the case of the Gaussian kernel, the multivariate kernel is defined as:

$$\mathcal{N}(\mathbf{x}; \boldsymbol{\mu}, \boldsymbol{\Sigma}) \triangleq \frac{1}{(2\pi)^{d/2} |\boldsymbol{\Sigma}|^{1/2}} \exp\left(-\frac{1}{2}(\mathbf{x} - \boldsymbol{\mu})^T \boldsymbol{\Sigma}^{-1}(\mathbf{x} - \boldsymbol{\mu})\right). \tag{6.20}$$

In the following, the notation $\mathcal{N}(\mathbf{x}; \boldsymbol{\mu}, \boldsymbol{\Sigma})$ is used to denote the d-dimensional Gaussian kernel defined in Equation (6.20).

Kernel bandwidth

The kernel bandwidth controls the region of influence of each training value. Essentially, the bandwidth controls the resolution of the density estimate – large smoothing parameters capture the global structure of the density whereas smaller values result in increased details and possibly artifacts. Consider the following two extreme cases that illustrate this.

- As the bandwidth approaches zero, the region of influence of each training value becomes infinitely narrow and the density estimate approaches a training delta function. Mathematically,

$$\sigma_{\mathbf{r}} \to 0, \qquad K\left(\frac{\mathbf{r} - \mathbf{r}_i(t)}{\sigma_{\mathbf{r}}}\right) \to \delta(\|\mathbf{r} - \mathbf{r}_i(t)\|), \tag{6.21}$$

where $\delta(\cdot)$ is the Kronecker delta function. In this extreme case, the density estimate significantly over-fits the training data.

- As the bandwidth becomes infinitely large, the region of influence of each training data also grows infinitely large. In this case, the density estimate approaches a uniform. Mathematically,

$$\sigma_{\mathbf{r}} \to \infty, \qquad K\left(\frac{\mathbf{r} - \mathbf{r}_i(t)}{\sigma_{\mathbf{r}}}\right) \to 1. \tag{6.22}$$

The above choice of the kernel width grossly under-fits the training data.

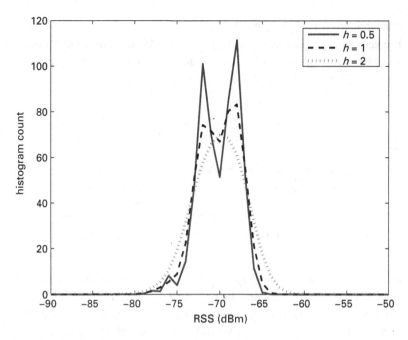

Figure 6.4. Effect of the smoothing parameter on the density estimate.

Figure 6.4 demonstrates the effect of the smoothing parameter h on the density estimate. Unfortunately, the accuracy of the density estimate is highly dependent on this parameter [75].

The choice of the kernel bandwidth is data-dependent and must be made such that

$$n\sigma_{\mathbf{r}} \to \infty \text{ as } n \to \infty. \tag{6.23}$$

Several theoretical approaches have been used to determine the *optimal* kernel bandwidth that minimizes the difference between the estimated and actual densities. For the Gaussian kernel, the optimal bandwidth is as follows [75]:

$$\sigma^* = \left(\frac{4}{d+2}\right)^{1/(d+4)} \hat{\sigma} n^{-1/(d+4)}. \tag{6.24}$$

In this equation, n, d, and $\hat{\sigma}$ are the number of samples, the dimension of the measurement vector, and the estimated standard deviation of the data, respectively. Other automatic methods, such as cross-validation, are also available for determination of the bandwidth.

Ideally, the kernel bandwidth should be small (and thus less smoothing) in areas where the density is large and large (resulting in more smoothing) where density is small. This leads to the development of adaptive choices of this parameter. These methods are beyond the scope of this book, and the interested reader is referred to [75] for a detailed account of the plethora of bandwidth selection techniques.

Product kernel density estimation

The kernel density estimator can also be used to estimate the joint density $f(\mathbf{p}, \mathbf{r})$. This is achieved through the *product kernel density estimator* [75]. Given a set of training pairs $\{(\mathbf{p}_i, \bar{\mathbf{r}}_i) | i = 1 \ldots N\}$, this density estimate is defined as

$$f(\mathbf{r}, \mathbf{p}) \approx \frac{1}{N \sigma_{\mathbf{r}} \sigma_{\mathbf{p}}} \sum_{i=1}^{N} K\left(\frac{\mathbf{r} - \bar{\mathbf{r}}_i}{\sigma_{\mathbf{r}}}\right) K\left(\frac{\mathbf{p} - \mathbf{p}_i}{\sigma_{\mathbf{p}}}\right), \qquad (6.25)$$

where the bandwidths $\sigma_{\mathbf{r}}$ and $\sigma_{\mathbf{p}}$ are determined according to Equation (6.24). Using the Gaussian kernel as before, the density estimate becomes

$$f(\mathbf{r}, \mathbf{p}) \approx \frac{1}{N} \sum_{i=1}^{N} \mathcal{N}(\mathbf{r}; \bar{\mathbf{r}}_i, \Sigma_{\mathbf{r}}) \mathcal{N}(\mathbf{p}; \mathbf{p}_i, \Sigma_{\mathbf{p}}), \qquad (6.26)$$

where $\Sigma_{\mathbf{r}} = \sigma_{\mathbf{r}}^* \mathbf{I}_d$ and $\Sigma_{\mathbf{p}} = \sigma_{\mathbf{p}}^* \mathbf{I}_d$. Note that this corresponds to the use of the multivariate Gaussian kernel with a diagonal kernel width matrix.

The second element in the training pair, $\bar{\mathbf{r}}_i$, is a single RSS value for each anchor point. Recall that at each anchor point several training samples are generally collected. This training record must somehow be collapsed into a *single* representative training value for use with the product kernel density estimator. Note that this approach is different from the likelihood density estimation technique where the *entire* RSS record was used at each anchor.

The RSS representative value at each anchor point must "best" represent the fingerprint record for the anchor point according to some criterion. Mathematically, the representative value is chosen to minimize an aggregate distance to RSS samples comprising the fingerprint. Therefore, we have

$$\bar{\mathbf{r}}_i = \arg\min_{\mathbf{r}} d(\mathbf{r}, \mathbf{F}(\mathbf{p}_i)), \qquad (6.27)$$

where $d(\mathbf{r}, \mathbf{F}(\mathbf{p}_i))$ is the aggregate distance or a loss function. An example of such loss function is the average square error given by

$$d_{\text{SE}}(\mathbf{r}, \mathbf{F}(\mathbf{p}_i)) = \sum_{t=1}^{n} (\mathbf{r} - \mathbf{r}_i(t))^T (\mathbf{r} - \mathbf{r}_i(t)). \qquad (6.28)$$

To find the representative value that minimizes the above distance, we differentiate the above function and set the derivative to zero. We then see that the loss function is minimized when

$$\bar{\mathbf{r}}_i = \frac{1}{n} \sum_{t=1}^{n} \mathbf{r}_i(t). \qquad (6.29)$$

That is, the RSS value that best represents the training data in the l_2 error norm sense is the mean of the training measurements.

We can also use the above approach to find the RSS representative value that minimizes the average l_1 norm as the loss function:

$$d_{AE}(\mathbf{r}, \mathbf{F}(\mathbf{p}_i)) = \sum_{t=1}^{n} |\mathbf{r} - \mathbf{r}_i(t))|. \tag{6.30}$$

This leads to the conclusion that [65]

$$\bar{\mathbf{r}}_i = median(\mathbf{r}_i(1), \ldots, \mathbf{r}_i(n)). \tag{6.31}$$

For computational simplicity, the mean value is generally used as the representative value in WLAN positioning systems.

6.2.3 Cautionary notes

Two cautionary notes are made in using the histogram and kernel density estimates. First, RSS measurements at each anchor location are generally collected over a short time window because of practical limitations (e.g. labor cost). Due to time variations in the propagation environment, however, RSS is not stationary over long time intervals [43], and a density estimate from a short-time window may not be suitable for use at a later time.

The second cautionary note relates to the assumption made by both the histogram and the KDE that the training samples are independent and identically distributed. This assumption is not true in the case of RSS measurements where correlation over intervals of several seconds are reported [43, 93].

Despite these shortcomings, both the histogram and kernel density estimates have been extensively and effectively used in WLAN positioning.

6.3 Memoryless position estimators

Section 6.2 discussed the theoretical fundamentals behind histogram and kernel density estimation. In this section, we use these techniques to develop maximum likelihood and minimum mean square error estimators.

6.3.1 Maximum likelihood estimation

As previously mentioned, the maximum likelihood position estimate is computed as follows:

$$\hat{\mathbf{p}}_{ML} = \arg\max_{\mathbf{p}_i} \hat{f}(\mathbf{r}|\mathbf{p}_i). \tag{6.32}$$

It now becomes clear that the maximum likelihood position estimate can be obtained by using either the histogram or the kernel density estimates to compute $\hat{f}(\mathbf{r}|\mathbf{p}_i)$.

6.3.2   Minimum mean square error estimate

Recall that the minimum mean square error (MMSE) estimate is the mean of the posterior density $f(\mathbf{p}|\mathbf{r})$. This density can be computed either from the likelihood density or from the joint density. We show in the following how each of these estimates can be used to develop position estimators.

Likelihood density

Using Bayes theorem, we have

$$f(\mathbf{p}|\mathbf{r}) = \frac{f(\mathbf{r}|\mathbf{p})f(\mathbf{p})}{\int_{\mathcal{P}} f(\mathbf{r}|\mathbf{p})f(\mathbf{p})d\mathbf{p}}. \tag{6.33}$$

Therefore, to obtain the posterior density, the prior density $f(\mathbf{p})$ and the likelihood density $f(\mathbf{r}|\mathbf{p})$ are needed.

The prior density reflects the knowledge of the environment before any RSS measurements are obtained. In the case of fingerprinting, the only thing known about the environment is the location of the anchor points. Given this information, a simple approximation to the prior density $f(\mathbf{p})$ is given by the empirical probability density function (epdf) [75],

$$f(\mathbf{p}) \approx \frac{1}{N} \sum_{i=1}^{N} \delta(\|\mathbf{p} - \mathbf{p}_i\|), \tag{6.34}$$

where $\delta(\cdot)$ denotes the Dirac delta function.[1]

Note that the anchor points are not always uniformly dispersed in the environment. In particular, the possible locations for these points are often determined by the layout of walls, furniture, and other obstructing objects. As such, the above density estimate provides coarse information regarding the topology of the environment. Substituting Equation (6.34) into Equation (6.33), we have

$$f(\mathbf{p}|\mathbf{r}) \approx \frac{f(\mathbf{r}|\mathbf{p})\sum_{i=1}^{N} \delta(\mathbf{p} - \mathbf{p}_i)}{\int_{\mathcal{P}} f(\mathbf{r}|\mathbf{p})\sum_{i=1}^{N} \delta(\|\mathbf{p} - \mathbf{p}_i\|)d\mathbf{p}} \tag{6.35}$$

$$= \frac{\sum_{i=1}^{N} f(\mathbf{r}|\mathbf{p}_i)\delta(\|\mathbf{p} - \mathbf{p}_i\|)}{\sum_{i=1}^{N} f(\mathbf{r}|\mathbf{p}_i)}. \tag{6.36}$$

Using this approximation to the posterior density, the posterior mean (MMSE estimate) is computed as follows:

$$\hat{\mathbf{p}} \approx \int_{\mathcal{P}} \mathbf{p} \frac{\sum_{i=1}^{N} f(\mathbf{r}|\mathbf{p}_i)\delta(\|\mathbf{p} - \mathbf{p}_i\|)}{\sum_{i=1}^{N} f(\mathbf{r}|\mathbf{p}_i)}d\mathbf{p}$$

---

[1] The epdf is the derivative of the staircase approximation to the cumulative probability distribution function constructed using the data.

$$= \frac{\sum_{i=1}^{N} \int_{\mathcal{P}} \mathbf{p} f(\mathbf{r}|\mathbf{p}_i) \delta(||\mathbf{p} - \mathbf{p}_i||) d\mathbf{p}}{\sum_{i=1}^{N} f(\mathbf{r}|\mathbf{p}_i)}$$

$$= \frac{\sum_{i=1}^{N} \mathbf{p}_i f(\mathbf{r}|\mathbf{p}_i)}{\sum_{i=1}^{N} f(\mathbf{r}|\mathbf{p}_i)}$$

$$= \sum_{i=1}^{N} \frac{w_i(\mathbf{r})}{\sum_{i=1}^{N} w_i(\mathbf{r})} \mathbf{p}_i, \tag{6.37}$$

where the weights are defined as

$$w_i(\mathbf{r}) \triangleq f(\mathbf{r}|\mathbf{p}_i). \tag{6.38}$$

The MMSE estimation problem is now reduced to estimating the likelihood density $f(\mathbf{r}|\mathbf{p}_i)$ from the location fingerprints. The likelihood density can be estimated using either the histogram or the kernel density estimator. That is

$$\hat{\mathbf{p}} = \sum_{i=1}^{N} \frac{\hat{f}(\mathbf{r}|\mathbf{p}_i)}{\sum_{i=1}^{N} \hat{f}(\mathbf{r}|\mathbf{p}_i)} \mathbf{p}_i. \tag{6.39}$$

Therefore, the MMSE position estimate is a linear combination of the anchor point coordinates with the weights determined by the likelihood density.

The MMSE estimation method used in this section uses anchor point information to estimate the prior density. This means that topological knowledge regarding the environment is implicitly incorporated into the design. However, as previously mentioned, this only provides a coarse approximation. An important limitation of this approximation is the assumption that the prior probability of the user occupying positions that do not correspond to anchor points is zero. Section 6.4 introduces an alternative design in which no assumptions on the prior density are used in the derivation of the MMSE estimate.

Joint density

An alternative approach taken to estimate the posterior density is to estimate this density directly from the joint density $f(\mathbf{p}, \mathbf{r})$. Recall that

$$f(\mathbf{p}|\mathbf{r}) = \frac{f(\mathbf{r}, \mathbf{p})}{\int_{\mathcal{P}} f(\mathbf{r}, \mathbf{p}) d\mathbf{p}}. \tag{6.40}$$

Substituting the product kernel density estimate of $f(\mathbf{p}, \mathbf{r})$ given in Equation (6.26) into Equation (6.40), the following estimate of the posterior density is obtained:

$$f(\mathbf{p}|\mathbf{r}) = \frac{f(\mathbf{p}, \mathbf{r})}{f(\mathbf{r})} \tag{6.41}$$

$$\approx \frac{\sum_{i=1}^{N} \mathcal{N}(\mathbf{r}; \bar{\mathbf{r}}_i, \boldsymbol{\Sigma}_{\mathbf{r}}) \mathcal{N}(\mathbf{p}; \mathbf{p}_i, \boldsymbol{\Sigma}_{\mathbf{p}})}{\sum_{i=1}^{N} \mathcal{N}(\mathbf{r}; \bar{\mathbf{r}}_i, \boldsymbol{\Sigma}_{\mathbf{r}})} \tag{6.42}$$

$$= \sum_{i=1}^{N} w_i(\mathbf{r}) \mathcal{N}(\mathbf{p}; \mathbf{p}_i, \boldsymbol{\Sigma}_{\mathbf{p}}), \tag{6.43}$$

where

$$w_i(\mathbf{r}) = \frac{\mathcal{N}(\mathbf{r}; \bar{\mathbf{r}}_i, \boldsymbol{\Sigma}_\mathbf{r})}{\sum_{i=1}^{N} \mathcal{N}(\mathbf{r}; \bar{\mathbf{r}}_i, \boldsymbol{\Sigma}_\mathbf{r})}. \tag{6.44}$$

The MMSE estimate and its associated covariance are the first two moments of the Gaussian mixture in Equation (6.43). These moments are [6]

$$\hat{\mathbf{p}} = \mathbb{E}\{\mathbf{p}|\mathbf{r}\} = \sum_{i=1}^{N} w_i(\mathbf{r}, F(\mathbf{p}_i))\mathbf{p}_i; \tag{6.45}$$

$$\mathbf{P} = \mathbb{E}\{(\mathbf{p} - \hat{\mathbf{p}})(\mathbf{p} - \hat{\mathbf{p}})^T|\mathbf{r}\} = \sum_{i=1}^{N} w_i(\mathbf{r}, F(\mathbf{p}_i))\left(\boldsymbol{\Sigma}_\mathbf{p} + (\mathbf{p}_i - \hat{\mathbf{p}})(\mathbf{p}_i - \hat{\mathbf{p}})^T\right). \tag{6.46}$$

In the above formulation, the posterior distribution is approximated using the spatially distributed set of anchor points and their representative RSS values. This estimator corresponds to the non-parametric Nadaraya–Watson regression estimator of the posterior mean $\mathbb{E}(\mathbf{p}|\mathbf{r})$ [64, 75]. Note, however, that the Nadaraya–Watson estimator cannot be directly applied to the positioning problem. Instead, a pre-processing step is needed to extract the representative RSS values for each anchor point. As we shall see in Chapter 7, an important benefit of using this estimator is the availability of the estimation covariance given in Equation (6.46).

6.4   Comments on the estimators

6.4.1   Summary

The three estimators discussed in this section are summarized in Table 6.1.

6.4.2   Complexity

The three estimators presented here differ greatly in complexity. In the case of the estimators relying on the likelihood density, the entire RSS record at each anchor point is used. In contrast, the estimator employing the joint density uses only a single representative RSS value per point to capture time variations in RSS. This has important implications on the complexity. Table 6.2 compares the computational and storage complexity of the above methods. Calculations assume off-line computation of observation-independent parameters including histograms, sample means and variances, and kernel bandwidths. The O notation is used here to provide an upper bound on the number of addition, multiplication, exponentiation, and sort operations needed for each algorithm. The storage complexity is reported as the number of floats and integers that need to be stored for positioning.

Note that the complexity of all algorithms is directly proportional to the number of access points and anchor points used in the estimation. Likelihood density estimation

Table 6.1. Summary of memoryless position estimators. Note that $\hat{f}(\mathbf{r}|\mathbf{p}_i)$ is estimated using a histogram or the kernel density estimator and $w_i(\mathbf{r}) = \frac{\mathcal{N}(\mathbf{r};\overline{\mathbf{r}}_i,\Sigma_\mathbf{r})}{\sum_{i=1}^N \mathcal{N}(\mathbf{r};\overline{\mathbf{r}}_i,\Sigma_\mathbf{r})}$

Maximum likelihood	$\hat{\mathbf{p}}_{ML} = \arg\max_{\mathbf{p}_i} \hat{f}(\mathbf{r}|\mathbf{p}_i)$
MMSE (likelihood)	$\hat{\mathbf{p}} = \sum_{i=1}^N \frac{\hat{f}(\mathbf{r}|\mathbf{p}_i)}{\sum_{i=1}^N \hat{f}(\mathbf{r}|\mathbf{p}_i)}\mathbf{p}_i$
MMSE (joint)	$\hat{\mathbf{p}} = \sum_{i=1}^N w_i(\mathbf{r}, F(\mathbf{p}_i))\mathbf{p}_i$
	$\mathbf{P} = \sum_{i=1}^N w_i(\mathbf{r}, F(\mathbf{p}_i)) \left(\Sigma_\mathbf{p} + (\mathbf{p}_i - \hat{\mathbf{p}})(\mathbf{p}_i - \hat{\mathbf{p}})^T\right)$

Table 6.2. Computational and storage complexity of the proposed methods. Parameters L, N, n, d, and b are the number of access points, the number of anchor points, the number of time samples per anchor point, and the number of histogram bins

	Computations				Storage	
	additions	multiplications	exps	sorts	integers	floats
Maximum likelihood (histogram)	$O(bdN)$	$O(dN)$	0	$O(N)$	0	$O(bLN)$
MMSE (likelihood) (KDE)	$O(ndN)$	$O(ndnN)$	$O(nN)$	0	$O(nLN)$	$O(N)$
MMSE (joint)	$O(dN)$	$O(dN)$	$O(N)$	0	0	$O(LN)$

LDE: likelihood density estimation, JDE: joint density estimation.

incurs the highest computational complexity since the entire RSS record is used to compute the weight for each anchor point.

6.4.3   Practical considerations

Bandwidth estimation

For the likelihood density estimation method, the bandwidth minimizing the positioning error is often larger than that predicted by Equation (6.24). This is due to two reasons. First, the non-stationarity of the RSS propagation environment means the density estimate based on the training data and bandwidth of Equation (6.24) does not sufficiently represent RSS measurements observed during the actual operation of the system. This problem is partially remedied by over-smoothing of the density estimate. The second contributing factor is the assumption that the RSS training samples used in the density and bandwidth estimation are independent and identically distributed. Since RSS measurements are correlated over short windows of time in practice, the variance used in the calculation of the bandwidth parameter is under-estimated.

For the joint density estimation method, the bandwidth value minimizing the positioning error is often smaller than that predicted by Equation (6.24). This happens as the bandwidth estimate of Equation (6.24) is known to lead to over-smoothing, especially for multimodal and skewed distributions [78].

One way to determine the applicability of the bandwidth predicted by Equation (6.24) is through validation. This is an off-line procedure whereby the positioning error for a known set of mobile positions is computed for different choices of bandwidth. The bandwidth with the lowest error is selected for the actual operation of the system. It is essential that the validation data set be different than the training set (radio map) to take into account the time variations in the environment.

Number of training samples

In general, the histogram estimate has a higher sensitivity to the number of training samples used as compared to the kernel density estimates. Theoretically, the rate of convergence of the histogram density estimator in the asymptotic mean integrated square error sense is $O(n^{-2/d+2})$, which is slower than the $O(n^{-4/4+d})$ rate achieved by the kernel density estimator. The joint density estimation method is the least sensitive to the number of training samples. In fact, an effective RSS representative can be obtained with as few as five RSS samples.

6.5   Experimental examples

6.5.1   Proximity-based positioning

The K-nearest-neighbor estimator defined in Equation (6.2) is evaluated using the two data sets discussed in Chapter 5. Figure 6.5 shows the positioning accuracy of this method as a function of the number of anchor points used in estimation. The figure shows the accuracy for both validation and testing data.

Three observations are made from these results.

(1) Including more than one anchor point improves positioning accuracy. In the case when only one anchor point is included, a noisy RSS observation may result in the inclusion of a sub-optimal anchor point. Averaging over multiple points increases the resiliency of the method to such outliers.

(2) The performance of this estimator is sensitive to the number of anchor points used in estimation. Because the contribution of all anchor points is equally weighed, including too many anchor points actually degrades the performance of this estimator. Validation data are effective in predicting the optimal number of anchor points (K) in the case of the stationary device (data set 1), but not for the moving device (data set 2).

(3) The estimator performs poorly for the second data set where the mobile device is moving. This is expected as the motion of the mobile device increases the variance of RSS measurements. This, in turn, increases the probability that the observed RSS deviates from the fingerprints stored in the radio map. The consequence of such a discrepancy is the inclusion of inappropriate anchor points in the average and a degradation in positioning accuracy.

6.5.2   Histogram-based positioning

Figure 6.6 shows the performance of the MMSE estimator (Equation (6.39)) using a histogram density estimate as a function of the histogram bin width. Three observations are in order.

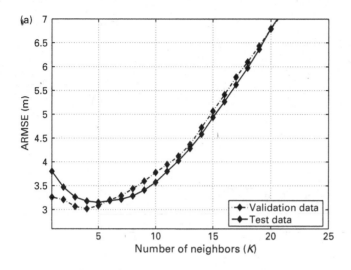

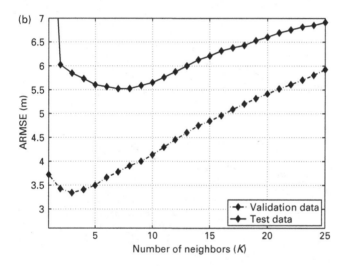

Figure 6.5.   Positioning accuracy of the K-nearest-neighbor estimator as a function of the number of anchor points used in the estimation. (a) Stationary user. (b) Mobile user.

(1) The performance of the histogram-based method is sensitive to the histogram bin width.

(2) The validation data are effective in predicting the optimal histogram bin width for both data sets.

(3) As seen for the K-nearest-neighbor estimator, the accuracy of the histogram-based estimator degrades significantly when the mobile device moves during positioning.

Figure 6.7 shows the positioning accuracy of the histogram-based method as a function of the number of anchor points included in the estimation. As before, including multiple

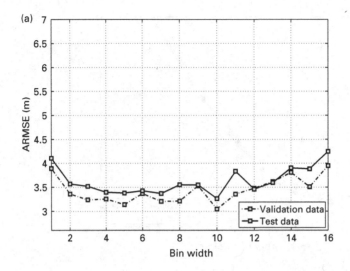

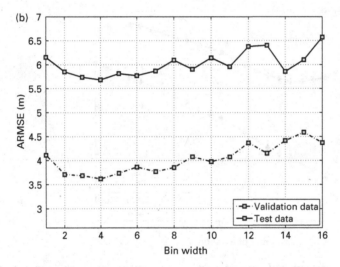

Figure 6.6. Positioning accuracy for the histogram-based MMSE estimator as a function of the histogram bandwidth. (a) Stationary user. (b) Mobile user.

anchor points does in fact improve positioning accuracy. In contrast to the K-nearest-neighbor estimator, however, increasing the number of anchor points does not degrade positioning accuracy. This is due to the weighting scheme used in Equation (6.39). In particular, this weighting scheme adjusts the contribution of each anchor point to the estimate based on the likelihood probability. Therefore, anchor points closest to the position of the mobile device receive a higher weight, whereas the contribution of anchor points far from the mobile device become negligible. These results clearly illustrate the benefits of including a weighting scheme during estimation.

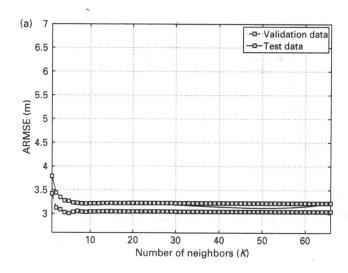

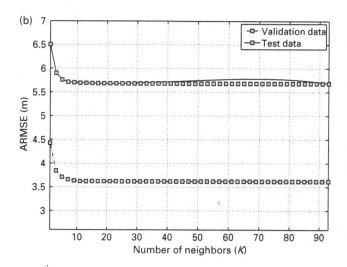

Figure 6.7.   Effect of number of neighbors (K) on the performance of the histogram-based method. (a) Stationary user. (b) Mobile user.

One disadvantage of the weighting scheme is the additional computational complexity introduced by including all the anchor points in the calculation. Fortunately, this overhead is minimal. In addition, the validation data are effective in predicting the optimal number of anchor points.

6.5.3   Kernel-density-estimate-based positioning

In this section, we explore the effect of the kernel bandwidth and the RSS representative value on the kernel-based positioning methods discussed in the previous sections.

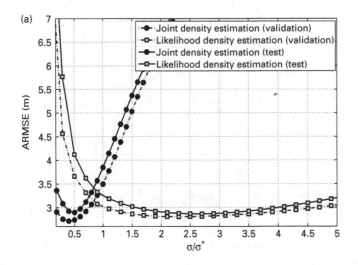

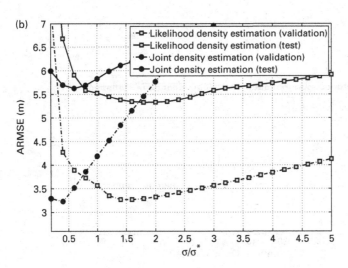

Figure 6.8.   Effect of kernel bandwidth on ARMSE. The value σ is the bandwidth used for estimation and σ^* is the bandwidth predicted by Equation (6.24). (a) Stationary user. (b) Mobile user.

Kernel bandwidth

Figure 6.8 shows the performance of both the likelihood and joint kernel methods as a function of the kernel bandwidth σ. Recall that Equation (6.24) predicts an optimal bandwidth, σ^*, for each estimator. In Figure 6.8, we investigate the optimality of this value by plotting the performance of the estimator versus multiples of σ^*.

As expected, the performance of the kernel methods is sensitive to the kernel bandwidth. Moreover, we see that the joint density estimator is more sensitive to the choice of this parameter than the likelihood density estimator. A second observation is that the bandwidth estimate provided by Equation (6.24) is not optimal for either estimator. In particular, we see that for the joint density estimator, the bandwidth predicted by

Equation (6.24) *over-fits* the data (too large) as predicted by [78]. In contrast, for the likelihood estimator, the predicted bandwidth *under-fits* the data (too small). This may be in part due to the correlation between various time samples used by this estimator. Fortunately, the validation data can predict the multiplier needed to achieve the highest positioning accuracy for both estimators.

As was the case with KNN and histogram estimators, we see that the positioning accuracy of the kernel density estimators degrade significantly with movement of the mobile device.

RSS representative

As noted previously, the joint density estimator reduces the entire training record into a single representative value to obtain the density estimate. Figure 6.9 depicts the

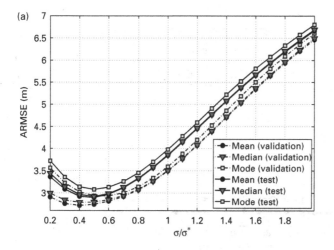

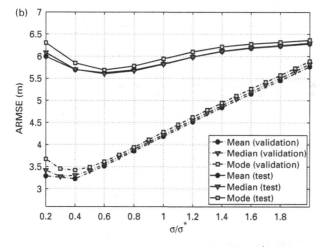

Figure 6.9. Effect of RSS representative choice on ARMSE (joint kernel density estimation). (a) Stationary user. (b) Mobile user.

performance of this estimator for the three methods of producing the representative value. These methods compute the representative value as the mean, median, and mode of the training sample, respectively.

The results indicate that the positioning accuracy is not highly sensitive to the choice of the representative value. In particular, the mean and median provide the highest accuracy, and they are closely followed by the mode. In light of these results, the mean value can be used as the RSS representative for computational simplicity.

6.5.4   Number of training points

Density estimation generally requires a large number of training samples. In this light, we investigate the performance of the estimators discussed in Section 6.4 for different sample sizes. The results are provided in Figure 6.10.

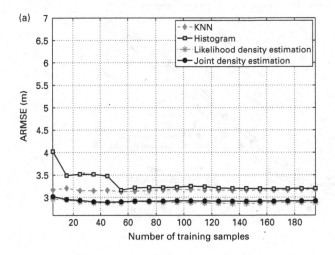

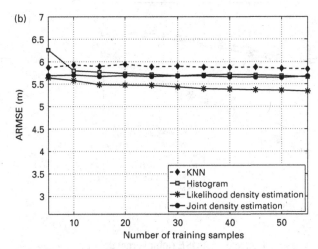

Figure 6.10.   Effect of the number of training samples on ARMSE. (a) Stationary user. (b) Mobile user.

From these results, we note that the histogram estimator is most sensitive to the number of training samples used. Theoretically, the rate at which the histogram density estimate converges to the true density is $O(n^{-2/d+2})$, where n is the number of time samples and d is the dimension of the measurement vector [75]. For the kernel density estimator, this rate is $O(n^{-4/4+d})$, which is faster than that of the histogram estimate. The KNN estimator does not rely on an estimate of the entire density, but employs only the first moment of this density. As such, this estimator requires the fewest number of training samples.

6.6   Chapter summary

We began this chapter by providing a mathematical formulation of the memoryless positioning problem, in which a single RSS observation and the radio map information are used to estimate the position of a mobile device. In particular, we considered two optimality criteria for positioning: (1) maximization of the likelihood probability density function, and (2) minimization of the mean square positioning error. We showed that position estimation using both of these criteria reduces to the problem of density estimation from the radio map. Since the parametric form of these densities is generally unknown in WLAN positioning, we presented two non-parametric density estimation techniques (histogram and kernel density estimation). Using these techniques, we then developed three position estimators that provide varying degrees of complexity and accuracy.

We concluded the chapter by evaluating the positioning estimators using experimental data. These results suggested that the positioning accuracy of all estimators degraded significantly when the user was moving. This may be attributed to the increased variations in RSS and deviations from the radio map data due to motion of the receiver. Chapter 7 will focus specifically on the development of techniques to address the challenges of this scenario; in particular, the knowledge of pedestrian movement dynamics as an additional information source during positioning to augment RSS measurements.

7 Model-based positioning

The previous chapter presented non-parametric solutions for memoryless MMSE estimation based on a single RSS observation at each time step. However, in most situations, additional RSS measurement noise resulting from the movement of the user poses a significant challenge to memoryless estimation. This chapter focuses on improving positioning accuracy through dynamic positioning or *tracking*. This paradigm exploits the correlation in positions observed over time resulting from the physical laws of kinematics. This is achieved by using the current RSS measurement as well as the measurement history and information from past position estimates.

The chapter opens with a description of the tracking problem and a discussion of the benefits of recursive MMSE estimation using Bayesian filtering (Sections 7.1 and 7.2). It then proceeds to introduce three special cases of the Bayesian filter, namely the Kalman filter (Sections 7.3 and 7.4), the non-parametric information filter (Section 7.5), and particle filters (Section 7.6) for RSS-based tracking in indoor WLANs. These filters provide a recursive position estimate over time by fusing information from RSS measurements and a motion model. We close this chapter by comparing the operation of these filters using experimental data (Section 7.7).

7.1 Bayesian tracking problem

In this chapter, we discuss estimators that continuously determine a sequence of minimum mean square estimates of position over time, $\hat{\mathbf{p}}(1), \ldots, \hat{\mathbf{p}}(k)$, given a sequence of noisy RSS measurements denoted as

$$\mathbf{R}(k) \triangleq \{\mathbf{r}_0, \ldots, \mathbf{r}(k)\}. \tag{7.1}$$

Note that, in contrast to memoryless positioning, the entire observation record is used for tracking [65].

As discussed in Chapter 6, the MMSE estimate of $\mathbf{p}(k)$ is the expected value of the position dependent on the observations. The MMSE position estimate is denoted as $\hat{\mathbf{p}}(k|j)$, where the notation $\hat{\mathbf{p}}(k|j)$ denotes the state estimate at time k given the observation record up to and including the jth observation ($\mathbf{R}(j)$). With this notation, the position estimate is given by

$$\hat{\mathbf{p}}(k|k) = \mathbb{E}\{\mathbf{p}(k)|\mathbf{R}(k)\} = \int_{\mathcal{P}} \mathbf{p}(k)f(\mathbf{p}(k)|\mathbf{R}(k))\mathrm{d}\mathbf{p}(k).$$

Evaluation of the above equation requires knowledge of the posterior density $f(\mathbf{p}(k)|\mathbf{R}(k))$. In the statistical sense, this density contains all the information about the past and current states of the system needed to form $\hat{\mathbf{p}}(k|k)$. Therefore, every time a new RSS measurement is received, the density $f(\mathbf{p}(k)|\mathbf{R}(k))$ must be determined to estimate the new position. Fortunately, this density can be computed *recursively*, eliminating the need for storage of the entire RSS measurement record. Using the recursive formulation, only the density $f(\mathbf{p}(k)|\mathbf{R}(k-1))$ and the new RSS measurement $\mathbf{r}(k)$ are needed. This storage and computational efficiency is of paramount importance in mobile positioning, where positioning computations are generally carried out on battery-operated devices with limited power and storage resources.

Recursive computation of the posterior density is accomplished using the Bayes filter, which relies on the Bayes theorem:

$$\begin{aligned}
f(\mathbf{x}(k)|\mathbf{R}(k)) &= f(\mathbf{x}(k)|\mathbf{r}(k),\mathbf{R}(k-1)) \\
&= \frac{f(\mathbf{r}(k)|\mathbf{x}(k),\mathbf{R}(k-1))}{f(\mathbf{r}(k)|\mathbf{R}(k-1))}f(\mathbf{x}(k)|\mathbf{R}(k-1)) \\
&= \frac{f(\mathbf{r}(k)|\mathbf{x}(k))}{f(\mathbf{r}(k)|\mathbf{R}(k-1))}f(\mathbf{x}(k)|\mathbf{R}(k-1)) \\
&= \underbrace{f(\mathbf{x}(k)|\mathbf{R}(k-1))}_{\text{model contribution}} \underbrace{\frac{f(\mathbf{x}(k)|\mathbf{r}(k))}{f(\mathbf{r}(k)|\mathbf{R}(k-1))}}_{\text{RSS contribution}} \underbrace{\frac{f(\mathbf{r}(k))}{f(\mathbf{x}(k))}}_{\text{prior knowledge}}.
\end{aligned} \tag{7.2}$$

In the above equation, $\mathbf{x}(k)$ is a *state vector* that contains all the variables needed to describe the position of the pedestrian carrying the mobile device, as we will see later. In the above derivation, we have assumed that the observation $\mathbf{r}(k)$ depends on $\mathbf{x}(k)$ but is conditionally independent from prior observations, i.e.,

$$f(\mathbf{r}(k)|\mathbf{x}(k),\mathcal{R}(k-1)) = f(\mathbf{r}(k)|\mathbf{x}(k)). \tag{7.3}$$

This assumption cannot be trivially justified because $\mathbf{r}(k)$ depends on environmental conditions contributing to multipath and shadowing in addition to the current position of the mobile. Despite this, as we will see, Bayesian filtering can significantly improve the accuracy of positioning.

According to Equation (7.2), the determination of the posterior density $f(\mathbf{x}(k)|\mathbf{R}(k))$ requires knowledge of four components, namely a prediction density, a likelihood density, measurement and measurement evolution densities, and a prior density.

Prediction density

The *prediction* density is denoted as $f(\mathbf{x}(k)|\mathbf{R}(k-1))$. As we shall see in Section 7.2.1, this density models the evolution of the pedestrian's motion over time by relating the observation record to the future position of the user. This is done by employing knowledge of the motion dynamics of the pedestrian. Essentially, this density accounts for the fact that the motion of pedestrians is correlated over time.

Likelihood density

The density $f(\mathbf{r}(k)|\mathbf{x}(k))$ is the likelihood density as before. This density relates the observed measurement to the position of the pedestrian. This density models the RSS–position relationship and is obtained from the radio map.

Measurement evolution density

The density $f(\mathbf{r}(k)|\mathbf{R}(k-1))$ models the time evolution of RSS. This density can be obtained from the prediction and likelihood densities since

$$f(\mathbf{r}(k)|\mathbf{R}(k-1)) = \int f(\mathbf{r}(k)|\mathbf{x}(k)) f(\mathbf{x}(k)|\mathbf{R}(k-1)) \mathrm{d}\mathbf{x}(k). \tag{7.4}$$

Measurement density

The density $f(\mathbf{r}(k))$ represents the distribution of RSS values. This density is dependent on how the RSS is actually reported by the network interface card. Specifically, this density is defined over the range of possible values of the reported RSS values. For simplicity, a uniform density defined over the range of RSS values reported by the wireless card is assumed.

Prior density

The prior density $f(\mathbf{x}(k))$ represents any information available about the position of the pedestrian before any observations are made. As discussed in Chapter 6, different formulations of the prior density can lead to different position estimators. In this chapter, we assume a uniform density over the space for computational convenience.

7.2   Predictor–corrector structure

Using the above formulation, the posterior density $f(\mathbf{x}(k)|\mathbf{R}(k))$ can be computed in two steps. First, a prediction density is computed based on the previous state of the system. Second, the incoming measurements are used to refine (correct) this prediction. Mathematically, these stages are stated as follows:

$$\textbf{prediction:} \quad f(\mathbf{x}(k-1)|\mathbf{R}(k-1)) \rightarrow f(\mathbf{x}(k)|\mathbf{R}(k-1)) \tag{7.5}$$

$$\textbf{state update (correction):} \quad f(\mathbf{x}(k)|\mathbf{R}(k-1)) \rightarrow f(\mathbf{x}(k)|\mathbf{R}(k)). \tag{7.6}$$

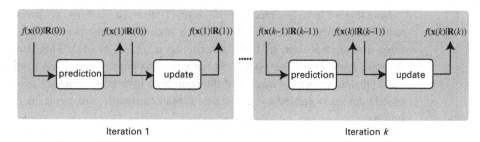

Figure 7.1.   Predictor–corrector structure.

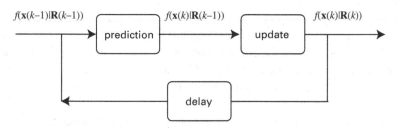

Figure 7.2.   Recursive implementation of the predictor–corrector structure.

Figure 7.1 illustrates the prediction and update steps involved in estimating the density $f(\mathbf{x}(k)|\mathbf{R}(k))$. As discussed in Section 7.2.1, the prediction step uses the knowledge of motion dynamics to form a *prediction* of the user's position. Once the RSS is received, this prediction is further corrected to improve accuracy and reduce the uncertainty associated with the estimate. We shall see in Section 7.2.2 that the correction or update step employs the knowledge of the likelihood density.

A key advantage of the predictor–corrector formulation is that this structure can be implemented recursively, thus reducing computational and storage needs. This recursive implementation is shown in Figure 7.2.

An important observation from Figure 7.1 is that the predictor–corrector formulation requires the initial density $f(\mathbf{x}(0)|\mathbf{R}(0))$. This issue will be further discussed in the following sections.

7.2.1   Prediction

The prediction step transforms the density of the position information at time $k - 1$ into a prediction at time k. This prediction is formed solely based on the motion dynamics and does not use the RSS information. To see this, consider the reformulation of the prediction density using the Chapman–Kolmogorov equation:

$$f(\mathbf{x}(k)|\mathbf{R}(k-1)) = \int f(\mathbf{x}(k)|\mathbf{x}(k-1))f(\mathbf{x}(k-1)|\mathbf{R}(k-1))\mathrm{d}\mathbf{x}(k-1). \quad (7.7)$$

Note that we have assumed a known initial state and a Markov-I property for the system. The density $f(\mathbf{x}(k-1)|\mathbf{R}(k-1))$ is known from the previous iteration of the filter, and all that remains to be determined is $f(\mathbf{x}|\mathbf{x}(k-1))$. This density defines the time evolution of the system. In our case, this density is derived from a *dynamic model* that describes the motion of the pedestrian over time. Here, we assume that the mobile device is carried by a pedestrian. Mathematically, this model is expressed as follows:

$$\mathbf{x}(k) = m(\mathbf{x}(k-1), \mathbf{w}(k)), \tag{7.8}$$

where $m(\cdot)$ is a linear or non-linear function and $\mathbf{w}(k)$ is the *system noise* representing modeling uncertainties.

As previously mentioned, $\mathbf{x}(k)$ is a *state vector* that contains the minimal set of parameters needed to describe the dynamic behavior of the system [33]. In tracking problems, the state vector contains parameters needed to describe the kinematics of the tracked target. In the case of WLAN tracking, the dynamic model describes the kinematics of the pedestrian movement over time.

Designing the dynamic model is generally non-trivial. This is because pedestrian motion dynamics are complex and governed by decision models, choice of destination, and interactions with other people or objects in the environment [2]. To reduce modeling complexities, simplifying assumptions are often used. For example, it may be assumed that in indoor office areas, pedestrian motion is constrained by the highly structured nature of hallways and rooms. Typical motion scenarios, for example, may include moving between offices and elevators through well-structured hallways. These types of motion are adequately described by a linear motion model that provides the displacement over a given time interval based on the velocity of the pedestrian. In this light, we define the state vector as [12, 17, 30, 88]

$$\mathbf{x}(k) \triangleq [p^x(k)\ v^x(k)\ p^y(k)\ v^y(k)]^T, \tag{7.9}$$

where $(p^x(k), p^y(k))$ and $(v^x(k), v^y(k))$ are the position and velocity of the pedestrian carrying the mobile at time k in two-dimensional Cartesian coordinates, respectively. From here two-dimensional, the objective is to obtain $\hat{\mathbf{x}}(k|j) \triangleq \mathbb{E}\{\mathbf{x}(k)|\mathbf{R}(j)\}$, the MMSE estimate of the state at time k given the observation record $\mathbf{R}(j)$. The position estimate is related to the state estimate as follows:

$$\hat{\mathbf{p}}(k|k) = \mathbf{A}\hat{\mathbf{x}}(k|k), \tag{7.10}$$

where

$$\mathbf{A} \triangleq \begin{pmatrix} 1 & 0 & 0 & 0 \\ 0 & 0 & 1 & 0 \end{pmatrix}. \tag{7.11}$$

The pedestrian model is inherently a continuous-time process. We define the continuous time state vector as $\mathbf{x}_c(t) \triangleq [p_c^x(t)\ v_c^x(t)\ p_c^y(t)\ v_c^y(t)]^T$. The simplified motion model used here is the second-order kinematics model [6, 65]

$$\frac{\mathrm{d}}{\mathrm{d}t}\mathbf{x}_c(t) = \mathbf{F}_c\mathbf{x}_c(t) + \mathbf{G}\mathbf{a}(t), \tag{7.12}$$

where

$$\mathbf{F}_c = \begin{bmatrix} 0 & 1 & 0 & 0 \\ 0 & 0 & 0 & 0 \\ 0 & 0 & 0 & 1 \\ 0 & 0 & 0 & 0 \end{bmatrix} \text{ and } \mathbf{G} = \begin{bmatrix} 0 & 0 \\ 1 & 0 \\ 0 & 0 \\ 0 & 1 \end{bmatrix}. \tag{7.13}$$

The vector $\mathbf{a}(t) = [a^x(t)\ a^y(t)]^T$ is a zero-mean continuous-time acceleration noise process with covariance given by

$$\mathbb{E}\{\mathbf{a}(t)\mathbf{a}^T(s)\} = q\delta(t-s), \tag{7.14}$$

where q is a design parameter representing the acceleration noise intensity. The choice of this parameter is guided by noting that changes in velocity over a period of length Δ are of the order $\sqrt{\Delta q}$ [6].

Due to the finite precision of digital computers implementing the positioning algorithms, as well as the discrete-time nature of the RSS measurements, this continuous-time model is discretized. Following the discretization procedure in [6, 65], the familiar linear-Gaussian model used in [12, 17, 30, 88] is obtained:

$$\mathbf{x}(k) = \mathbf{F}\mathbf{x}(k-1) + \mathbf{w}(k), \tag{7.15}$$

where $\mathbf{w}(k) \sim \mathcal{N}(0, \mathbf{Q})$ is white Gaussian noise independent of the state $\mathbf{x}(k)$. The system matrix is given by

$$\mathbf{F} = e^{\mathbf{F}_c\Delta} = \begin{pmatrix} 1 & \Delta & 0 & 0 \\ 0 & 1 & 0 & 0 \\ 0 & 0 & 1 & \Delta \\ 0 & 0 & 0 & 1 \end{pmatrix}. \tag{7.16}$$

The parameter Δ is the discretization period, which is set to the sampling rate for the RSS measurements. In this work, $\Delta = 0.5$ seconds, the maximum rate provided by the measurement software.

The noise covariance $\mathbf{Q}$ matrix in the discrete-time model is determined from the continuous-time motion model as described in [65]:

$$\mathbf{Q} = q \begin{pmatrix} \Delta^3/3 & \Delta^2/2 & 0 & 0 \\ \Delta^2/2 & \Delta & 0 & 0 \\ 0 & 0 & \Delta^3/3 & \Delta^2/2 \\ 0 & 0 & \Delta^2/2 & \Delta \end{pmatrix}. \tag{7.17}$$

We now have a dynamic model that can be used to derive the prediction density $f(\mathbf{x}(k)|\mathbf{x}(k-1))$. Given this and the density $f(\mathbf{x}(k-1) - \mathbf{R}(k-1))$, Equation (7.7) is used to determine the prediction density $f(\mathbf{x}(k)|\mathbf{R}(k-1))$. Note that, depending on the form of $f(\mathbf{x}(k)|\mathbf{R}(k-1))$, Equation (7.7) may not have a closed-form solution. In fact, a closed-form solution can only be found under certain conditions, which we discuss in the following sections.

7.2.2   Update (correction)

Once the prediction step is completed and an RSS measurement is received, the update or correction step is carried out to refine the position estimate further. The objective of the update step is therefore to find the density $f(\mathbf{r}(k)|\mathbf{x}(k))$ given $f(\mathbf{x}(k)|\mathbf{R}(k-1))$ and the RSS measurement. Similar to the prediction step, the density $f(\mathbf{r}(k)|\mathbf{x}(k))$ is obtained using a model relating the observations to the state. The general model for observations is stated mathematically as follows:

$$\mathbf{r}(k) = g(\mathbf{x}(k), \mathbf{v}(k)), \tag{7.18}$$

where $g(\cdot)$ is a linear or non-linear function and $\mathbf{v}(k)$ is a noise term representing modeling uncertainties. Note that $\mathbf{x}(k)$ is the unobservable state of the system. The above model is known as the *measurement model* and relates RSS measurements to the position of the pedestrian. As discussed in Chapter 6, determination of this model is one of the most challenging aspects of designing a WLAN positioning system.

Equations (7.8) and (7.18) are referred to as the state-space equations. Using these equations, the prediction and update stages provide the solution for propagating and updating the state estimate from one time step to the next. Unfortunately, calculation and propagation of the posterior density is intractable only in a few specific cases. As a consequence, suboptimal approaches to the Bayes filter must be considered. The remaining part of this chapter reviews some of these approaches.

7.3   Kalman filter

The *Kalman filter* was developed by Rudolf E. Kalman in 1960 [45]. Since then, it has become one of most widely used implementations of the Bayesian filter. This filter is the closed-form solution of Equation (7.2) when both the system and the measurement equations are linear and the noise processes are Gaussian. In other words, the Kalman filter provides the optimal MMSE estimate of the pedestrian's position under linear-Gaussian conditions.

In Sections 7.4 and 7.5, we discuss the implementation of this filter and the extension to the WLAN positioning problem.

7.3.1   Prediction

The prediction step of the Kalman filter uses a linear-Gaussian system model to compute Equation (7.7). In particular, assuming the linear-Gaussian pedestrian motion model discussed previously, we see that

$$f(\mathbf{x}(k)|\mathbf{x}(k-1)) = \mathcal{N}(\mathbf{F}\mathbf{x}(k-1), \mathbf{Q}).\qquad(7.19)$$

Substituting this into Equation (7.7), we have

$$f(\mathbf{x}(k)|\mathbf{R}(k-1)) = \int f(\mathbf{x}(k)|\mathbf{x}(k-1))f(\mathbf{x}(k-1)|\mathbf{R}(k-1))\mathrm{d}\mathbf{x}(k-1)\qquad(7.20)$$

$$= \int \mathcal{N}(\mathbf{F}\mathbf{x}(k-1), \mathbf{Q})\mathcal{N}(\hat{\mathbf{x}}(k-1), \mathbf{P}(k|k-1))\mathrm{d}\mathbf{x}(k-1)\qquad(7.21)$$

Here we have made use of an important assumption underlying the Kalman filter: the posterior density $f(\mathbf{x}(k)|\mathbf{R}(k))$ is Gaussian for all k. This follows from the assumption of the linear-Gaussian system and measurement models. In particular, we have assumed that

$$f(\mathbf{x}(k)|\mathbf{R}(k)) = \mathcal{N}(\hat{\mathbf{x}}(k), \mathbf{P}(k|k)),\qquad(7.22)$$

where $\hat{\mathbf{x}}(k)$ is the state estimate at time k (recall that the MMSE estimate is the posterior mean) and $\mathbf{P}(k|k)$ is the covariance associated with the estimate.

After computing the integral in Equation (7.21), the prediction density $f(\mathbf{x}(k)|\mathbf{R}(k-1))$ becomes a Gaussian density $\mathcal{N}(\hat{\mathbf{x}}(k|k-1), \mathbf{P}(k|k-1))$, with

$$\hat{\mathbf{x}}(k|k-1) = \mathbf{F}\hat{\mathbf{x}}(k-1|k-1),\qquad(7.23)$$

$$\mathbf{P}(k|k-1) = \mathbf{F}\mathbf{P}(k-1|k-1)\mathbf{F}^T + \mathbf{Q}.\qquad(7.24)$$

7.3.2 Update (correction)

Given the prediction density computed above, the update step must be carried out. To do this, the Kalman filter assumes a linear-Gaussian measurement model:

$$\mathbf{r}(k) = \mathbf{H}\mathbf{x}(k) + \mathbf{v}(k).\qquad(7.25)$$

Using this model and the prediction density, the closed-form solution of the Bayes filter (Equation (7.2)) can be found. In particular, the posterior density is computed to be $f(\mathbf{x}(k)|\mathbf{R}(k)) = \mathcal{N}(\hat{\mathbf{x}}(k), \mathbf{P}(k|k))$, where

$$\hat{\mathbf{x}}(k|k) = \hat{\mathbf{x}}(k|k-1) + \mathbf{K}\left(\mathbf{r}(k) - \mathbf{H}\hat{\mathbf{x}}(k|k-1)\right),\qquad(7.26)$$

$$\mathbf{P}(k|k) = (1 - \mathbf{K}\mathbf{H})\mathbf{P}(k|k-1)(1 - \mathbf{K}\mathbf{H})^T + \mathbf{K}\mathbf{P}_\mathbf{r}\mathbf{K}^T.\qquad(7.27)$$

The value $\mathbf{K}$ is known as the *Kalman gain* and is given by

$$\mathbf{K} = \mathbf{P}(k|k-1)\mathbf{H}\left[\mathbf{H}\mathbf{P}(k|k-1)\mathbf{H}^T + \mathbf{P}_\mathbf{r}\right]^{-1}.\qquad(7.28)$$

The Kalman algorithm is outlined in Algorithm 7.1 for reference.

Algorithm 7.1. Outline of the Kalman filter algorithm

Assumptions
- state-space equation:
$$\mathbf{x}(k+1) = \mathbf{F}\mathbf{x}(k) + \mathbf{w}(k),$$
$$\mathbf{r}(k) = \mathbf{H}\mathbf{x}(k) + \mathbf{v}(k),$$
where $\mathbf{F}$ is the state transition matrix.
- $\mathbf{w}(k)$, $\mathbf{v}(k)$ are uncorrelated, zero-mean, Gaussian white noise sequences with
$$E\mathbf{w}(k)\mathbf{w}(j)^T = \delta_{kj}\mathbf{Q}, \; E\mathbf{v}(k)\mathbf{v}(j)^T = \delta_{kj}\mathbf{P_r}, \; E\mathbf{v}(k)\mathbf{w}(j)^T = 0.$$
- Initial state is a Gaussian random variable $\mathbf{x}(0)$ with
$$E\mathbf{x}(0) = \hat{\mathbf{x}}(0|0) \text{ and } E(\mathbf{x}(0) - \hat{\mathbf{x}}(0))(\mathbf{x}(0) - \hat{\mathbf{x}}(0))^T = \mathbf{P}(0|0).$$
- $\mathbf{v}(k)$, $\mathbf{w}(k)$, and initial state are uncorrelated.

Prediction
$$\hat{\mathbf{x}}(k|k-1) = \mathbf{F}\hat{\mathbf{x}}(k-1|k-1),$$
$$\mathbf{P}(k|k-1) = \mathbf{F}\mathbf{P}[k-1|k-1]\mathbf{F}^T + \mathbf{Q}.$$

Update
$$\hat{\mathbf{x}}(k|k) = \hat{\mathbf{x}}(k|k-1) + \mathbf{K}(\mathbf{r}(k) - \mathbf{H}\hat{\mathbf{x}}(k|k-1)),$$
$$\mathbf{P}(k|k) = (1 - \mathbf{KH})\mathbf{P}(k|k-1)(1 - \mathbf{KH})^T + \mathbf{K}\mathbf{P_r}\mathbf{K}^T,$$
where $\mathbf{K}$ is given by
$$\mathbf{K} = \mathbf{P}(k|k-1)\mathbf{H}[\mathbf{H}\mathbf{P}(k|k-1)\mathbf{H}^T + \mathbf{P_r}]^{-1}.$$

7.3.3 Comments

Interpretation

Under the conditions of a linear-Gaussian system and measurement models, the Kalman filter provides the optimal MMSE estimate given a set of observations. This is achieved by optimally combining two sources of information: knowledge of system dynamics and information obtained from measurements. This optimal combination is performed using the Kalman gain. Intuitively, this gain regulates the contribution of each of the two sources of information to the final estimate (see Equation (7.26)). Note specifically that the value of the Kalman gain is directly proportional to the prediction covariance $\mathbf{P}(k|k-1)$ and inversely proportional to the measurement covariance $\mathbf{P_r}$. The covariance matrix provides a measure of uncertainty associated with the estimate (larger covariance corresponds to larger uncertainty). As such, the Kalman gain regulates the contribution of each information source based on its associated uncertainty.

Limitations

The major shortcoming of the Kalman filter is the restrictive assumptions on the state-space model. Several methods have been proposed to deal with non-linearities in the models including the Extended [65] and Unscented [72] Kalman Filters. Direct application of these methods to the problem of WLAN positioning, however, is not possible. The main challenge in WLAN tracking is that the measurement equation, Equation

(7.18), is not explicitly known because of the complexity of the propagation environment. That is, both the function $g(\cdot,\cdot)$ and the statistical description of the noise process $\mathbf{v}(k)$ are unknown in Equation (7.18). Instead, training RSS values collected at a set of spatially distributed anchor points implicitly characterize the RSS–position dependency. The lack of an explicit relation between the state and the RSS measurements prevents the direct application of the Kalman filter and its variants in the WLAN tracking problem.

7.4   Modified Kalman filter

As previously mentioned, the application of the Kalman filter to the WLAN positioning problem is not possible as the measurement equation is unknown. One way to overcome this limitation is to build a *synthetic* measurement equation that replaces RSS measurements with entities whose relationship to the state vector (position) is known. As shown in Figure 7.3, this can be achieved by using a pre-processor to generate pseudo-measurements for the filter [62, 30, 51]. Essentially, the pre-processor is used to replace the measurement equation with one of the non-parametric memoryless position estimators discussed previously.

Instead of using the traditional measurement equation, which relates the state to the observed RSS, this design uses a synthetic measurement equation that relates the memoryless position estimate to the state. The measurement equation then takes the following form:

$$\hat{\mathbf{p}}(k) = \mathbf{H}\mathbf{x}(k) + \mathbf{v}(k), \tag{7.29}$$

where $\hat{\mathbf{p}}(k)$ is the position estimate produced by the memoryless estimator at time k, and

$$\mathbf{H} = \begin{bmatrix} 1 & 0 & 0 & 0 \\ 0 & 0 & 1 & 0 \end{bmatrix}. \tag{7.30}$$

The entity $\mathbf{v}(k)$ is the Gaussian noise associated with this estimate. This noise is assumed to have zero mean, but its covariance must be determined. This can be implemented in two ways.

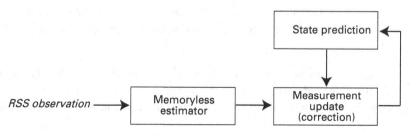

Figure 7.3.   Overview of the modified Kalman filter.

Algorithm 7.2. Outline of the modified Kalman filter algorithm

---

Inputs

    Anchor point information: $\{(\mathbf{p}_1, \bar{\mathbf{r}}_1), \ldots (\mathbf{p}_N, \bar{\mathbf{r}}_N)\}$,

    RSS observation at time k: $\mathbf{r}(k)$.

Outputs

    State estimate at time k: $\hat{\mathbf{x}}(k|k)$.

Assumptions

- State-space equation:

$$\mathbf{x}(k+1) = \mathbf{F}\mathbf{x}(k) + \mathbf{w}(k),$$
$$\hat{\mathbf{p}}(k) = \mathbf{H}\mathbf{x}(k) + \mathbf{v}(k),$$

    where $\mathbf{F}$ is the state transition matrix and $\hat{\mathbf{p}}(k)$ is a

    memoryless position estimate at time k.

- $\mathbf{w}(k)$, $\mathbf{v}(k)$ are uncorrelated, zero-mean, Gaussian

    white noise sequences with

$$E\mathbf{w}(k)\mathbf{w}(j)^T = \delta_{kj}\mathbf{Q}, \; E\mathbf{v}(k)\mathbf{v}(j)^T = \delta_{kj}\mathbf{P_r}, \; E\mathbf{v}(k)\mathbf{w}(j)^T = 0.$$

- Initial state is a Gaussian random variable $\mathbf{x}(0)$ with

$$E\mathbf{x}(0) = \hat{\mathbf{x}}(0|0) \text{ and } E(\mathbf{x}(0) - \hat{\mathbf{x}}(0))(\mathbf{x}(0) - \hat{\mathbf{x}}(0))^T = \mathbf{P}(0|0).$$

- $\mathbf{v}(k)$, $\mathbf{w}(k)$, and initial state are uncorrelated.

Prediction

$$\hat{\mathbf{x}}(k|k-1) = \mathbf{F}\hat{\mathbf{x}}(k-1|k-1),$$
$$\mathbf{P}(k|k-1) = \mathbf{F}\mathbf{P}[k-1|k-1]\mathbf{F}^T + \mathbf{Q}.$$

Update

$$\hat{\mathbf{x}}(k|k) = \hat{\mathbf{x}}(k|k-1) + \mathbf{K}(\mathbf{r}(k) - \mathbf{H}\hat{\mathbf{x}}(k|k-1)),$$
$$\mathbf{P}(k|k) = (1 - \mathbf{K}\mathbf{H})\mathbf{P}(k|k-1)(1 - \mathbf{K}\mathbf{H})^T + \mathbf{K}\mathbf{P_r}\mathbf{K}^T,$$

    where $\mathbf{K}$ is given by

$$\mathbf{K} = \mathbf{P}(k|k-1)\mathbf{H}\left[\mathbf{H}\mathbf{P}(k|k-1)\mathbf{H}^T + \mathbf{P_r}\right]^{-1}.$$

- The first method for determining the noise covariance in the above equation is to assume a constant covariance. This constant can be computed based on knowledge of the memoryless estimator in a given environment, or by using validation data.
- The second method for determining the noise covariance exploits the fact that the joint density memoryless estimator presented in Chapter 6 provides an estimated covariance as a byproduct of the positioning process. Therefore, if this estimator is used to find $\hat{\mathbf{p}}(k)$, the noise covariance will be computed at every time step (see Equation (6.46)).

Regardless of the method used to determine the noise covariance, the state vector and the estimated position are assumed to be linearly dependent in Equation (7.29). As such, the traditional formulation of the Kalman filter can be used to obtain the position estimate.

The modified Kalman algorithm is outlined in Algorthm 7.2 for reference.

7.4.1   Comments

Interpretation

The above formulation enables the application of the Kalman filter to the problem of WLAN positioning. Essentially, this formulation employs the Kalman gain to fuse the position estimate obtained from memoryless estimation with information obtained from a motion model.

Limitations

The modified Kalman filter replaces the unknown measurement equation with a synthetic equation. This, however, requires explicit assumptions on the measurement matrix $\mathbf{H}$ and the noise covariance $\mathbf{v}(k)$. Depending on the choice of the memoryless estimator to be used and the noise covariance, these assumptions may not always hold.

7.5   Non-parametric information filter

In contrast to the Kalman filter, where measurement contributions are obtained from the measurement equation, the measurement contribution (update) can be computed directly using the joint kernel density estimate (KDE) memoryless position estimator proposed in Chapter 6. This approach eliminates the need for the use of the synthetic measurement equation presented in Equation (7.29). Given the non-parametric estimation of position using the KDE, the closed-form solution of the Bayes filter is known as the non-parametric information (NI) filter [54]. The NI filter fuses information from this memoryless estimator with the prediction obtained from the dynamic model to handle effectively the non-linear and non-Gaussian conditions of the tracking problem. This section details the derivation of the NI filter.

7.5.1   Prediction

The prediction density for the NI filter is determined in a similar manner to that for the Kalman filter. In particular, the system equation modeling dynamics of pedestrian movements is used to derive the prediction density $f(\mathbf{x}(k)|\mathbf{x}(k-1))$. This density is then used with the Chapman–Kolmogorov equation and the assumption that $\mathbf{x}(k)$ is a Markov-I process to obtain the prediction density. As before,

$$f(\mathbf{x}(k)|\mathbf{x}(k-1)) = \mathcal{N}(\mathbf{F}\mathbf{x}(k-1), \mathbf{Q}). \tag{7.31}$$

Therefore, the prediction density $f(\mathbf{x}(k)|\mathbf{R}(k-1))$ is again a Gaussian density $\mathcal{N}(\hat{\mathbf{x}}(k|k-1), \mathbf{P}(k|k-1))$, with

$$\hat{\mathbf{x}}(k|k-1) = \mathbf{F}\hat{\mathbf{x}}(k-1|k-1), \tag{7.32}$$

$$\mathbf{P}(k|k-1) = \mathbf{F}\mathbf{P}(k-1|k-1)\mathbf{F}^T + \mathbf{Q}. \tag{7.33}$$

The density $f(\mathbf{x}(k-1)|\mathbf{R}(k-1))$ corresponds to the state estimate from the previous time step. An important difference between the Kalman and NI filters is that the Kalman filter assumes $f(\mathbf{x}(k-1)|\mathbf{R}(k-1))$ to be Gaussian, whereas the NI filter *approximates* this density with a Gaussian density. This is important as a non-linear measurement equation will lead to non-Gaussianity of $f(\mathbf{x}(k-1)|\mathbf{R}(k-1))$. The following sections explain how the NI filter approximates this density as a Gaussian.

7.5.2   Update

Recall that the second component needed to compute the posterior density $f(\mathbf{x}(k)|\mathbf{R}(k))$ is the measurement contribution $f(\mathbf{x}(k)|\mathbf{r}(k))$. In classical state-space formulations, such as the Kalman filter, this density is approximated using an explicit function relating RSS measurements to the state vector. In the case of WLAN positioning, however, such an explicit function describing the RSS–position relationship is unknown. Recall that this relationship is instead represented implicitly using the radio map. As a result, the measurement equation must be approximated from the radio map $\mathcal{R} = \{\mathbf{p}_i, \mathbf{F}(\mathbf{p}_i)\}_{i=1}^{N}$. Using the results of Chapter 6, the NI filter uses the joint KDE memoryless estimator for this purpose.

Recall that the KDE requires a set of training pairs $\{(\mathbf{x}_i, \bar{\mathbf{r}}_i)|i=1,\dots,N\}$ that characterize joint RSS position density. In Chapter 6, the joint density estimator employed the training pairs $\{(\mathbf{p}_i, \bar{\mathbf{r}}_i)|i=1,\dots,N\}$ obtained from the radio map. For the state-space formulation developed here, in addition to the x and y coordinates describing the position of the anchor points, the state vector also includes velocity information. Therefore, before applying the KDE, state training values $\mathbf{x}_i$ must be generated from the anchor points $\mathbf{p}_i = [p_i^x \ p_i^y]^T$ by augmenting this information with missing velocity values. Since the radio map does not provide any knowledge regarding the mobile's motion, the velocity values (and acceleration values when applicable) are set to zero and $\mathbf{x}_i = [p_i^x \ 0 \ p_i^y \ 0]^T$.

Given this set of training pairs, $\{(\mathbf{x}_i, \bar{\mathbf{r}}_i)|i=1,\dots,N\}$, and assuming a Gaussian kernel as before, the kernel density estimate is given by

$$f(\mathbf{x}(k)|\mathbf{r}(k)) \approx \sum_{i=1}^{N} w_i(\mathbf{r}(k)) \mathcal{N}(\mathbf{x}(k); \mathbf{x}_i, \boldsymbol{\Sigma}_{\mathbf{x}}), \tag{7.34}$$

with weights obtained as

$$w_i(\mathbf{r}(k)) = \frac{\mathcal{N}(\mathbf{r}(k); \bar{\mathbf{r}}_i, \boldsymbol{\Sigma}_{\mathbf{r}})}{\sum_{i=1}^{N} \mathcal{N}(\mathbf{r}(k); \bar{\mathbf{r}}_i, \boldsymbol{\Sigma}_{\mathbf{r}})}. \tag{7.35}$$

The diagonal matrices $\boldsymbol{\Sigma}_{\mathbf{r}} = \sigma_{\mathbf{r}}^2 \mathbf{I}$ and $\boldsymbol{\Sigma}_{\mathbf{x}}$ correspond to the covariance of the Gaussian kernel function. The parameter $\sigma_{\mathbf{r}}$ and diagonal elements of $\boldsymbol{\Sigma}_{\mathbf{x}}$ corresponding to position are chosen according to Equation (6.24). Since the fingerprints do not offer any motion information, velocity components of $\boldsymbol{\Sigma}_{\mathbf{x}}$ are determined using the typical pedestrian velocity values (for example, 1 meter/second).

In the above formulation, the posterior distribution is determined using a Gaussian mixture obtained from a spatially distributed set of points. Essentially, this approximation replaces the measurement equation used in the Kalman filter. The MMSE state estimate and its covariance determined based on RSS measurements only are the first two moments of the Gaussian mixture in Equation (7.34), as discussed in Chapter 6:

$$\hat{\mathbf{x}}_{\mathbf{r}}(k) = \mathbb{E}\{\mathbf{x}(k)|\mathbf{r}(k)\} = \sum_{i=1}^{N} w_i(\mathbf{r}(k))\mathbf{x}_i, \tag{7.36}$$

$$\mathbf{P}_{\mathbf{r}}(k) = \mathbb{E}\{(\mathbf{x}(k) - \hat{\mathbf{x}}(k))(\mathbf{x}(k) - \hat{\mathbf{x}}(k))^T|\mathbf{r}(k)\}$$

$$= \sum_{i=1}^{N} w_i(\mathbf{r}(k)) \left(\mathbf{\Sigma}_{\mathbf{x}} + (\mathbf{x}_i - \hat{\mathbf{x}}(k))(\mathbf{x}_i - \hat{\mathbf{x}}(k))^T \right). \tag{7.37}$$

Substituting the densities $\mathcal{N}(\hat{\mathbf{x}}(k|k-1), \mathbf{P}(k|k-1)), \mathcal{N}(\hat{\mathbf{x}}_{\mathbf{r}}(k), \mathbf{P}_{\mathbf{r}}(k))$ into Equation (7.2), the posterior density is obtained as follows:

$$f(\mathbf{x}(k)|\mathbf{R}(k)) = f(\mathbf{x}(k)|\mathbf{R}(k-1)) \frac{f(\mathbf{x}(k)|\mathbf{r}(k))}{f(\mathbf{r}(k)|\mathbf{R}(k-1))} \frac{f(\mathbf{r}(k))}{f(\mathbf{x}(k))}$$

$$\approx \mathcal{N}(\hat{\mathbf{x}}(k|k), \mathbf{P}(k|k)).$$

The MMSE state estimate $\hat{\mathbf{x}}(k|k) = \mathbb{E}\{\mathbf{x}(k)|\mathbf{R}(k)\}$ and its covariance are given by [54]

$$\hat{\mathbf{x}}(k|k) = \mathbf{P}(k|k) \left(\mathbf{P}^{-1}(k|k-1)\hat{\mathbf{x}}(k|k-1) + \mathbf{P}_{\mathbf{r}}^{-1}\hat{\mathbf{x}}_{\mathbf{r}}(k) \right), \tag{7.38}$$

$$\mathbf{P}^{-1}(k|k) = \mathbf{P}^{-1}(k|k-1) + \mathbf{P}_{\mathbf{r}}^{-1}(k). \tag{7.39}$$

The final components needed for the recursive computation of the MMSE estimates are the initial conditions $\hat{\mathbf{x}}(0|0)$ and $\mathbf{P}(0|0)$. Since no information is available on the position of the mobile device before measurements are made, initial conditions are set to initial measurement estimates, i.e. $\hat{\mathbf{x}}(0|0) = \hat{\mathbf{x}}_{\mathbf{r}}(0)$ and $\mathbf{P}(0|0) = \mathbf{P}_{\mathbf{r}}(0)$.

This completes the prediction and update steps, which comprise one iteration of the NI filter. The complete filter algorithm is shown in Algorithm 7.3.

Figure 7.4 graphically depicts the operation of the NI filter.

7.5.3 Comments

Interpretation

The main novelty of the NI filter lies in the non-parametric formulation of the measurement update step. This step involves the linear fusion of model and measurement contributions, where the weight of each source is proportional to its respective inverse covariance. In the case of Gaussian distributions, these inverse covariances are especially significant. In particular, these inverse covariances represent the *Fisher information matrix* [65, 66]. The Fisher information matrix provides a measure of the amount of information, or conversely the uncertainty, associated with the state contained from each each source. Intuitively, the source that conveys more information about the state (lowest uncertainty) receives a higher weight during fusion.

Algorithm 7.3. Non-parametric information (NI) filter

Inputs
  Anchor point information: $\{(\mathbf{x}_1, \bar{\mathbf{r}}_1), \ldots (\mathbf{x}_N, \bar{\mathbf{r}}_N)\}$.
  RSS observation at time k: $\mathbf{r}(k)$.
Outputs
  State estimate at time k: $\hat{\mathbf{x}}(k|k)$.
  Estimation covariance at time k: $\mathbf{P}(k|k)$.
NI filter
  Model-based prediction
  $\hat{\mathbf{x}}(k|k-1) = \mathbf{F}\hat{\mathbf{x}}(k-1|k-1)$,
  $\mathbf{P}(k|k-1) = \mathbf{F}\mathbf{P}(k-1|k-1)\mathbf{F}^T + \mathbf{Q}$.
  Measurement update
  Spatial processing:
  $$w_i(\mathbf{r}(k)) = \mathcal{N}(\mathbf{r}(k); \mathbf{r}_i, \boldsymbol{\Sigma}_\mathbf{r}) \left(\sum_{i=1}^{N} w_i(\mathbf{r}(k))\right)^{-1},$$
  $$\hat{\mathbf{x}}_\mathbf{r}(k) = \sum_{i=1}^{N} w_i(\mathbf{r}(k))\mathbf{x}_i,$$
  $$\mathbf{P}_\mathbf{r}(k) = \sum_{i=1}^{N} w_i(\mathbf{r}(k)) \left(\boldsymbol{\Sigma}_\mathbf{x} + (\hat{\mathbf{x}}_\mathbf{r}(k) - \mathbf{x}_i)(\hat{\mathbf{x}}_\mathbf{r}(k) - \mathbf{x}_i)^T\right).$$
  State estimation:
  $$\mathbf{P}(k|k)^{-1}\hat{\mathbf{x}}(k|k) = \mathbf{P}(k|k-1)^{-1}\hat{\mathbf{x}}(k|k-1) + \mathbf{P}_\mathbf{r}^{-1}\hat{\mathbf{x}}_\mathbf{r}(k),$$
  $$\mathbf{P}(k|k)^{-1} = \mathbf{P}^{-1}(k|k-1) + \mathbf{P}_\mathbf{r}^{-1}(k).$$
  Initial conditions
  $\hat{\mathbf{x}}(0|0) = \hat{\mathbf{x}}_\mathbf{r}(0)$, $\mathbf{P}(0|0) = \mathbf{P}_\mathbf{r}(0)$.

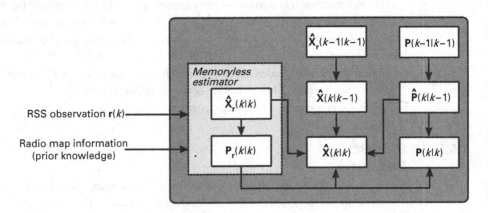

Figure 7.4.   Operation of the non-parametric information filter.

An important observation from the covariance update $\mathbf{P}(k|k)^{-1} = \mathbf{P}(k|k-1)^{-1} + \mathbf{P}_\mathbf{r}^{-1}(k)$ is that every iteration of the NI filter decreases the uncertainty in the state (increases observation). This formulation is similar to that of the *information filter* [36, 66, 61], an algebraic equivalent of the Kalman filter. The information filter uses a linear

Gaussian measurement equation to compute the measurement contributions, whereas the NI filter replaces the measurement equation with a non-parametric spatial processor based on anchor point and fingerprint information.

Limitations

The NI filter is suboptimal with respect to the MMSE criterion introduced in Equation (6.3) for several reasons. First, recall that the derivation of the NI filter assumes independence of the measurement noise. In particular, the derivation of Equation (7.2) assumed conditional independence of the measurements over time. This assumption is generally not true in WLANs as RSS measurements are highly dependent on environmental factors that affect radio propagation. These factors include doors opening or closing, elevators moving across floors, and users present in the area. Since these changes occur over time scales greater than the sampling period of the RSS, they introduce correlation into the measurements. Another factor that may result in correlated measurements is the systematic errors of the RSS measurement apparatus, as well as the kernel density estimator (e.g. bias [52]). Finally, this filter estimates the non-Gaussian posterior density with a single Gaussian density at every iteration. The Gaussian approximation of the measurement equation also results in suboptimality of the estimate $\hat{\mathbf{x}}(k|k)$ with respect to the MMSE criterion. The RSS measurements used in Equation (7.34) are not the true RSS values but quantized values provided by the network interface card. This leads to non-Gaussian quantization noise that is processed through the non-linear relationship of Equation (7.34). This type of processing generally leads to non-Gaussian posterior distributions.

7.6   Particle filter

The Kalman and NI filters provide a suboptimal solution to the Bayesian filtering problem by propagating the first two moments of the posterior density (mean and covariance) through time. As previously discussed, however, the Kalman filter relies on the linear-Gaussian assumption, which is often violated in practice due to the non-linear relationship between RSS measurements and position. The NI does not assume that the posterior density is Gaussian, but approximates the non-Gaussian density with a Gaussian density by computing the first two moments of this non-linear density at each iteration.

An alternative solution is to estimate the entire posterior density at each iteration. Unfortunately, this is impractical since an infinite number of points are needed to represent a continuous density. As an approximation to this approach, a discrete representation of the density can be used to reduce computation, storage and communication requirements. For example, a discrete and piecewise approximation of the posterior can be used by assuming that the state space is composed of P discrete possibilities $\{\mathbf{x}^1 \dots, \mathbf{x}^P\}$.

The prediction and update steps are then estimated as follows [72]:

$$f(\mathbf{x}|\mathbf{R}(k-1)) \approx \sum_{i=1}^{P} w_i(k|k-1)\delta(\mathbf{x}(k-1) - \mathbf{x}^i), \qquad (7.40)$$

$$f(\mathbf{x}|\mathbf{R}(k)) \approx \sum_{i=1}^{P} w_i(k|k)\delta(\mathbf{x} - \mathbf{x}^i). \qquad (7.41)$$

The weights represent the posterior probability $p(\mathbf{x} = \mathbf{x}^i|\mathbf{R})$ and are defined as follows:

$$w_i(k|k-1) \approx \sum_{j=1}^{P} w_j(k-1|k-1)p(\bar{\mathbf{x}}^i(k)|\bar{\mathbf{x}}^j(k-1)), \qquad (7.42)$$

$$w_i(k|k) \approx \frac{w_i(k|k-1)p(\mathbf{r}|\bar{\mathbf{x}}^i(k))}{\sum_{j=1}^{P} w_j(k|k-1)p(\mathbf{r}|\bar{\mathbf{x}}^j(k))}, \qquad (7.43)$$

where $\bar{\mathbf{x}}^j(k)$ denotes the center of the jth segment at time k. An important disadvantage of this grid-based method is that although the samples are chosen to represent the significant portion of the density globally, there is no way to concentrate the samples on the high probability regions to obtain variable resolution. This shortcoming is remedied in the *particle filtering* approach. Particle filters provide a recursive implementation of the Bayes filter using Monte Carlo techniques. Particle filtering is also referred to as bootstrap filtering, the condensation algorithm, sequential importance sampling, interacting particle approximations, and survival of the fittest [3] in the existing literature. The key idea is to propagate a set of samples of the probability density functions (as opposed to the actual density) at each update step.

Particle filters take advantage of Monte Carlo numerical integration methods to evaluate Equation (7.2). In these approaches, an integrand is factorized, where possible, as follows [72]:

$$\int g(x)\mathrm{d}x = \int f(x)\pi(x) \approx \frac{1}{P}\sum_{i=1}^{P} f(x_i), \qquad (7.44)$$

where $\pi(x)$ is a probability distribution and hence the approximation corresponds to a sample mean. Since it may be difficult to generate samples directly from the distribution $\pi(x)$, the *principle of importance sampling* allows this density to be replaced by another density, or *importance function*, $q(x)$, that has the same support as $\pi(x)$. Then [72],

$$\int g(x)\mathrm{d}x = \int f(x)\frac{\pi(x)}{q(x)}q(x) \approx \frac{1}{N}\sum_{i=1}^{P} w(x_i)f(x_i), \qquad (7.45)$$

with

$$w(x_i) = \frac{\pi(x_i)/q(x_i)}{\sum_{i=1}^{P} \pi(x_i)/q(x_i)}. \qquad (7.46)$$

Algorithm 7.4. Outline of the regularized particle filtering algorithm

---

Models

$$\mathbf{x}(k) = \mathbf{F}\mathbf{x}(k-1) + \mathbf{w}(k),$$

$$f(\mathbf{r}(k)|\mathbf{x}^j(k)) \approx \frac{\sum_{i=1}^{N} \mathcal{N}(\mathbf{r};\bar{\mathbf{r}}_i,\Sigma_{\mathbf{r}})\mathcal{N}\left(\mathbf{x}^j(k);\mathbf{x}_i,\Sigma_{\mathbf{x}}\right)}{\sum_{i=1}^{N} \mathcal{N}(\mathbf{x}^j(k);\mathbf{x}_i,\Sigma_{\mathbf{x}})}.$$

Prediction

Draw samples $\mathbf{x}^j(k) \sim f(\mathbf{x}(k)|\mathbf{x}^j(k-1))$, $j = 1,\dots,N_{\text{particles}}$.

Update

Calculate $\tilde{w}^j(k) \propto w^j(k-1)f(\mathbf{r}(k)|\mathbf{x}^j(k))$, $j = 1,\dots,N_{\text{particles}}$.

Normalize weights $w^j(k) = \frac{\tilde{w}^j(k)}{\sum_{j=1}^{N_{\text{particles}}}}$, $j = 1,\dots,N_{\text{particles}}$.

$\hat{\mathbf{x}}(k|k) = \sum_{j=1}^{N_{\text{particles}}} w^j(k)\mathbf{x}^j(k)$,

$\mathbf{P}(k|k) = \sum_{j=1}^{N_{\text{particles}}} w^j(k)(\mathbf{x}^j(k) - \hat{\mathbf{x}}(k|k))(\mathbf{x}^j(k) - \hat{\mathbf{x}}(k|k))^T$.

Calculate $N_{\text{eff}} = \frac{1}{\sum_{j=1}^{N_{\text{particles}}} (w^j(k))^2}$.

If $N_{\text{eff}} < N_{\text{thresh}}$, resample according to [3].

The principle of importance sampling is used to approximate the posterior distribution using a random measure[1] $\{\mathbf{x}^i, w^i\}_{i=1}^{P}$ as shown in Equation (7.47) [3]:

$$p(\mathbf{x}(k)|\mathbf{R}(k)) \approx \sum_{i=1}^{P} w^i(k)\delta(\mathbf{x}(k) - \mathbf{x}^i(k)), \tag{7.47}$$

and the weights are shown in Equation (7.48),

$$w_i(k) \propto w_i(k-1)\frac{p(\mathbf{r}|\mathbf{x}^i(k))p(\mathbf{x}^i(k)|\mathbf{x}^i(k-1))}{q(\mathbf{x}^i(k)|\mathbf{x}^i(k-1),\mathbf{r})}. \tag{7.48}$$

In this equation, $q(\mathbf{x}|\mathbf{r})$ is an importance density.

The particle filter suffers from a problem known as the *degeneracy phenomenon*. This refers to the fact that after a few iterations of the algorithm, all the particles except for one are assigned negligible weights [18]. The effects of this problem can be reduced by the choice of the importance sampling function and *resampling* [18]. Resampling aims to eliminate particles with small weights and focus on ones with larger weights. The various versions of the particle filters in the literature can be derived from the base algorithm discussed here by an appropriate choice of the importance sampling function and the resampling algorithm [3]. These parameters are discussed in great detail in [3, 16, 72]. Algorithm 7.4 provides an algorithmic view of the particle filter used here. The reader is referred to [3, 16, 72] for more detailed studies of the filter and its variations.

For the positioning problem, the particle filter can use the same linear-Gaussian motion model as the NI and Kalman filters. Moreover, a kernel density estimator is used for the

---

[1] For a definition of random measures, the reader is referred to [5].

estimation of the particle weights as shown in Algorithm 7.4. The resampling threshold can be chosen as $N_{\text{thresh}} = (2/3)N_{\text{particles}}$ as recommended in [29].

7.6.1 Comments

Interpretation

The Kalman and NI filters find suboptimal solutions to the Bayes filter based on the assumption that the posterior density can be estimated as Gaussian in each time step. The particle filter does not suffer from this limitation. Instead, it propagates the densities

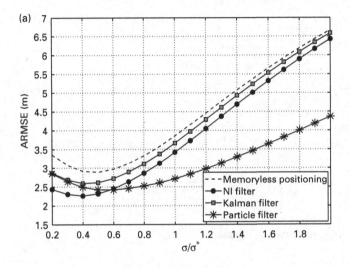

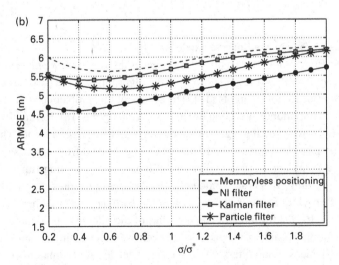

Figure 7.5.   Effect of the kernel bandwidth on the average root mean square error for the NI, Kalman, and particle filters. (a) Stationary user. (b) Mobile user.

Table 7.1. Positioning error statistics for the memoryless estimator compared to those of Kalman, particle, and NI filters (stationary user)

Method	ARMSE (m)	Mean (m)	Variance (m^2)
Memoryless KDE	2.90	2.44	4.98
Kalman filter	2.75	2.41	6.24
Particle filter	2.44	2.17	3.46
NI filter	2.29	2.08	3.00

through time using a set of samples. As such, this filter can capture the non-linear and non-Gaussian conditions inherent in the WLAN positioning problem.

Limitations

The main limitation of the particle filter relates to its computational complexity. In particular, the number of particles used for positioning is generally large (more than 1000 particles). This results in a high computational cost for the particle filtering, hindering its usefulness in realtime positioning applications on mobile devices with limited computational and power resources.

7.7 Experimental examples

To illustrate the concepts discussed in this chapter, Figure 7.5 depicts the performance accuracy of the Kalman, NI, and particle filters as a function of the kernel bandwidth used. For comparison, the positioning accuracy of the memoryless joint kernel density estimator discussed in the previous section is also provided.

The performance of all estimators is sensitive to the choice of the kernel bandwidth. Fortunately, the optimal bandwidth for model-based estimators can be chosen in a manner similar to that used for the memoryless estimator in Chapter 6.

All model-based estimators improve positioning accuracy as compared to the memoryless estimator. To further examine this, Tables 7.1 and 7.2 provide positioning accuracies of each estimator for the best value of the kernel bandwidth for stationary and mobile users. As seen, the model-based estimators are able to reduce the mean and variance of the positioning error as compared to memoryless estimation. The improvements are especially evident for the mobile user. This further illustrates the benefits of model-based estimation. As expected, the use of the motion model provides an additional source of information that can be used to combat the unpredictable nature of RSS disturbances.

Despite the improvements offered by model-based estimators, the positioning accuracy of all estimators degrades significantly in the case of mobile users. This is especially evident in the high variances associated with positioning error. As mentioned previously, for the mobile user, the movement of the device may introduce additional variance in RSS measurements. This further increases the chance that the on-line RSS observations

Table 7.2. Positioning error statistics for the memoryless estimator compared to those of Kalman, particle, and NI filters (mobile user)

Method	ARMSE (m)	Mean (m)	Variance (m^2)
Memoryless KDE	5.70	3.94	24.25
Kalman filter	5.41	3.78	22.05
Particle filter	5.16	3.74	20.63
NI filter	4.58	3.37	16.68

differ significantly from the RSS fingerprints of relevant anchor points, leading to a degradation in positioning accuracy.

7.8   Chapter summary

In this chapter, we introduced Bayesian filtering as a means to combat the detrimental effects of the noise associated with RSS measurement on positioning accuracy. Bayesian filtering uses the knowledge of pedestrian motion dynamics as an additional source of information during positioning. In particular, Bayesian filter fuse information from a model that describes the motion of the mobile device with information obtained from the RSS measurements and the radio map to improve positioning accuracy and consistency. We began by describing the optimal Bayesian filter. We then proceeded to discuss three suboptimal versions of this filter that are applicable to the WLAN positioning problem. These were the Kalman filter, the non-parametric information (NI) filter, and the particle filter.

In addition to improving positioning accuracy and consistency, the Bayesian filters discussed in this chapter predict future positions of the mobile device. Chapter 8 will discuss how these predictions can be used to select anchor points and access points proactively to improve positioning further.

8 Sensor selection

Positioning systems employ information from a large number of access points and anchor points distributed over large areas to locate a mobile device. In order to form a position estimate, the system must combine or *fuse* the information provided by these sources to make inferences regarding the location of a mobile device. Fusion of data from multiple sources is not a trivial task as redundancy and conflict among the information can significantly affect the accuracy and reliability of the final estimate. In the context of positioning, data fusion is especially challenging as the unreliable and time-varying nature of the radio-propagation channel means that erroneous and out-of-date information may be received from access and anchor points. In this Chapter, we focus on one data fusion challenge, namely sensor selection. In our context, sensor selection refers to selecting a subset of the available access and anchor points for positioning.

The rest of this chapter is organized as follows. We begin by motivating sensor selection for WLAN-based positioning (Section 8.1). We then proceed to discuss several methods for access point and anchor point selection (Sections 8.2 and 8.3). We conclude this chapter by illustrating the benefits of sensor selection using experimental data (Section 8.4).

8.1 Motivation

In WLAN-positioning, information from multiple anchor points and access points is fused to form a position estimate. Careful selection of the sensors that contribute to positioning can benefit RSS-based positioning in two ways.

- **Scalability** The complexity of positioning computation is directly proportional to the number of access and anchor points used. As the coverage area of the positioning system grows, so do the number of anchor and access points. As a result, increasing the area of service directly increases the computational complexity of positioning. To promote scalability, positioning computations can be carried out on spatially localized areas. In other words, positioning can be carried out using only the *relevant* access and anchor points.
- **Accuracy** As previously mentioned, the indoor propagation environment is highly time varying. This time variance may cause discrepancies between the radio map and on-line RSS observations. Such discrepancies complicate the positioning task

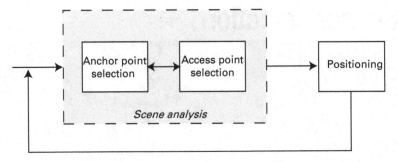

Figure 8.1.   Overview of scene analysis.

and degrade the performance of memoryless and dynamic estimators. This is seen in particular in the case of moving devices, as shown in previous chapters. Sensor selection can be used to lessen the undesirable effects of such uncertainties associated with RSS measurement. In particular, sensor selection can ensure that only anchor and access points that provide reliable information are used in positioning. Through effective sensor selection, a positioning system can dynamically adjust its view of the environment and deal with the unreliable nature of the radio propagation channel.

Figure 8.1 shows an example of a positioning system that incorporates both access point and anchor point selection methods. These functions are typically carried out after receiving an on-line observation and before positioning. Essentially, the sensor selection module provides the positioning system with a lens that filters out unnecessary information.

In this book, we refer to the collection of access points and anchor points as the *radio scene*. Sensor selection then entails analyzing the entire radio scene and restricting the positioning to a localized area of this scene. Sections 8.2 and 8.3 discuss access point and anchor point selection in detail.

8.2   Access point selection

Positioning in two-dimensional coordinates can be carried out with as few as three access points. In large indoor environments, such as shopping malls and hospitals, however, hundreds of access points may be available for use in positioning. Access point selection refers to the process of selection of a subset of all available access points for positioning. It is important to note that each WLAN access point has a limited coverage (typically less than 100 meters). As a result, as a mobile device moves in an environment, it will receive coverage from different sets of access points. For this reason, access point selection must be carried out every time a new RSS observation becomes available.

In mathematical terms, the objective of access selection is to determine a subset d from the L available access points. This selection is carried out such that the positioning obtains the "best" view of the environment at any time step based on a given optimality criterion. In this section, we discuss several optimality criteria and the resulting selection algorithms.

8.2.1   Strongest access point

The most common method for access point selection is to choose d access points with the strongest signal strength at the receiver. This method is advantageous in that the strongest access points provide the highest probability of coverage over time [90]. Moreover, selection of the strongest access point involves only a simple search and is therefore computationally inexpensive.

The strongest access point method suffers from two disadvantages. First, the variance of measurements from an access point generally increases with its mean power at a given location. A high RSS variance means that the observed RSS value may be very different than the RSS signatures stored in the radio map, leading to degradation in positioning accuracy [42]. Furthermore, it becomes more difficult to distinguish neighboring anchor points in such cases. The second disadvantage of this method is that selection of access points does not directly consider improvements in positioning accuracy. This point will become clear in the following sections.

8.2.2   Measure of diversity

The RSS value from a particular access point at a given location depends on the distance between the location and the access point as well as the geometry of the environment (for example, the layout of the walls and furniture). Depending on these factors, one or more access points may provide similar RSS values for a given location. Moreover, access points that are located close to each other are affected by the same channel impediments. In data fusion, however, it is important that the sensor readings contributing to a final estimate provide *complementary* information. This is critical for two reasons. First, complementarity ensures that the maximum amount of information is used in the estimation. Second, inclusion of sensors suffering from the same impediments will bias the estimation results. Motivated by these observations, one method for selection of access points chooses d access points such that the amount of information or diversity is maximized. In this section, we examine three examples of this method, namely information gain, Bhattacharyya distance, and principal/independent component analysis. These methods all aim to minimize redundancy between the selected access point, though they each use a different measure of redundancy.

Information gain

The first method uses the entropy-based information gain criterion [14] as the measure of diversity. This method essentially chooses the access point set that maximizes the amount of information provided to the positioning algorithm. This criterion is defined as follows:

$$InfoGain(a_n) = H(G) - H(G|a_n), \tag{8.1}$$

where $H(G)$ is the entropy of the radio map when a_n is excluded and $H(G|a_n)$ is the conditional entropy of the radio map given a_n. The information gain measure is advantageous in that it directly maximizes the information included in the chosen access points. However, since entropy computation relies on various probability distributions which must be estimated, this measure is suboptimal.

Bhattacharyya distance

Another method for measuring the diversity between access points is the Bhattacharyya distance. This measure operates based on the principle of minimizing the redundancy between selected access points. It evaluates the diversity between a pair of access points as the distance between the respective empirical RSS distributions obtained during training.

The Bhattacharyya distance [19] is a simple and computationally efficient method for evaluating the distance between two distributions. Mathematically, this distance is defined for two probability distributions $f_1(\mathbf{r})$ and $f_2(\mathbf{r})$ as follows:

$$d_{\text{Bhatt}} = -\ln\left(\int \sqrt{f_1(\mathbf{r})f_2(\mathbf{r})}d\mathbf{r}\right). \qquad (8.2)$$

A closed-form solution of the above integral exists in the case of two Gaussian distributions with known mean and covariance. Assuming the RSS distribution is Gaussian for two access points, this distance becomes

$$d_{\text{Bhatt}}(a_m, a_n|\mathbf{p}_i) = \frac{1}{8}(\bar{r}_i^m - \bar{r}_i^n)^2 \left(\frac{(\sigma_i^m)^2 + (\sigma_i^n)^2}{2}\right)^{-1} + \frac{1}{2}\ln\left(\frac{(\sigma_i^m) + (\sigma_i^n)}{2\sigma_i^m\sigma_i^n}\right), \qquad (8.3)$$

where $\bar{r}_i^n$ and σ_i^n are the sample mean and variance, respectively, of access point a_n at anchor point $\mathbf{p}_i$. The total distance between two access points is defined as the minimum divergence over anchor points:

$$D(a_n, a_m) = \min_{\mathbf{p}_i} d_{\text{Bhatt}}(a_m, a_n|\mathbf{p}_i). \qquad (8.4)$$

Finally, the total divergence for a set of d access points is computed as the sum of pairwise divergences.

The distance $d_{\text{Bhatt}}(a_m, a_n|\mathbf{p}_i)$ can be computed off-line, but the selection method still requires an exhaustive search over $\binom{L}{d}\binom{d}{2}$ sets. To reduce complexity and ensure that the selected access points provide coverage to anchor points near the observation point, access point selection is performed on the strongest $L' < L$ access points.

There are two issues of concern with the Bhattacharyya method. First, maximization of the Bhattacharyya distance between RSS distributions may not correspond to a selection of access points with the best physical geometry for positioning. Second, the method relies only on the first two moments of the RSS distribution at each anchor point. This is clearly suboptimal as these distributions are generally non-Gaussian.

Independent/principal component analysis

Traditionally, positioning is carried out in a vector space defined by the access points (recall that the RSS vector comprises readings from all access points in the environment). Due to possible redundancies discussed above, the access points defining this space may not be independent. In this light, a third example of access point selection transforms the vector space defined by the access points into another vector space whose components are independent. This transformation is achieved through the use of independent component analysis or principal component analysis [22]. Instead of using the original access points to carry out positioning, the independent components are used for this purpose.

8.2.3 Measure of discrimination ability

The measures discussed in Section 8.2.2 aimed to maximize the complementarity of the access points selected for positioning. This focus of this section is on measures that consider the ability of anchor points to discriminate between anchor points. In other words, these methods choose a set of access points that provide the best discrimination ability over space. In particular, we present two examples of selection criteria based on discrimination ability.

Fisher criterion

One way to quantify the discrimination ability of an access point is the Fisher criterion [19, 77]. This criterion relies on two premises for access point selection. The first premise is that we wish to choose access points for which the RSS values do not vary greatly over time. This ensures consistency between training and on-line values. The second desirable property for an access point is strong discrimination ability, evaluated as the strength of variations of mean RSS across anchor points. In particular, we wish to choose access points whose RSS values at different anchor points are greatly different.

Mathematically, the Fisher score for an access point a_n is defined as follows [54]:

$$\xi(a_n) = \frac{\sum_{i=1}^{N(k)} (\overline{r}_i^n - \overline{r}^n)^2}{\sum_{i=1}^{N(k)} \left(\sum_{t=1}^{n} (r_i^n(t) - \overline{r}_i^n)^2 \right)}, \tag{8.5}$$

where $\overline{r}_i^n = (1/n) \sum_{t=1}^{n} r_i^n(t)$ and $\overline{r}^n = (1/N(k)) \sum_{i=1}^{N(k)} \overline{r}_i^n$ are the sample mean of the training RSS values from access point a_n at $\mathbf{p}_i$ and across all anchor points, respectively. Access points with the highest scores $\xi(a_n)$ are chosen for positioning.

Fisher discriminant analysis

Recall our previous discussion regarding the use of a transformed space to carry out positioning (through independent component analysis or principal component analysis). It is possible to design the transformed space so that the ability to discriminate between anchor points is maximized in this space. This can be accomplished through the use of Fisher discriminant analysis [19, 21, 77].

8.3   Anchor point selection

The position estimates discussed in Sections 8.1 and 8.2 are generally a weighted average of anchor points. The weight of each anchor point is proportional to the distance between the observed RSS and the RSS fingerprint at that anchor point. Due to discrepancies between observations and fingerprint values resulting from the non-stationarity of propagation environment, anchor points far away from the observation point may erroneously receive high weights. Therefore, it is desirable to *localize* the positioning algorithm to a relevant area of the space. In other words, instead of including the entire anchor point set in the weighted sum that generates the position estimate, only a relevant subset of anchor points are included. In addition to increasing positioning accuracy, this method will also decrease the computational complexity of positioning. In this section, we focus on determination of a *region of interest* (ROI) that includes relevant anchor points at a given time.

8.3.1   RSS-based ROI

The first approach to determine the ROI relies on the assumption that anchor points that are close in proximity generally have similar RSS signatures. Therefore, upon observing an RSS vector during the on-line operation of the system, a set of anchor points with similar signatures is determined that comprises the ROI. Using this assumption, several techniques can be employed for anchor point selection.

One example of anchor point selection techniques involves an off-line clustering of locations aiming to reduce the search space to a single cluster which constitutes the ROI [14, 90]. Both of the above clustering techniques are carried out off-line based on the training data. This hampers the operation of the system over time since WLAN infrastructures are highly dynamic and access points can easily be moved or discarded. To overcome this limitation, the ROI can be determined on-line. This can be accomplished through on-line spatial filtering based on the premise that points that are close in the physical space receive coverage from similar sets of access points [52]. This method considers binary coverage vectors instead of actual RSS values, which are prone to greater time variations.

Recall that statistical characteristics of access points vary over space. The notion of the ROI, therefore, is also advantageous as it enables localized evaluation of access point selection criteria as discussed in Section 8.2.

8.3.2   Feedback-based ROI

The above ROI selection methods rely on RSS values that are susceptible to unpredictable variations due to time variations in the propagation channel. These methods are, in essence, memoryless: they only rely on one observation in time. As we saw previously, however, several benefits can be achieved if the previous history of the mobile's position is taken into account. This forms the basis of a feedback-based ROI determination technique [54]. This method uses the estimated or predicted position of the mobile device to determine the center of the ROI after each RSS observation is received.

The above technique is especially suitable for use with dynamic positioning algorithms such as those employing the Kalman filter, NI filter, or particle filter. In particular, the one-step position predictions from these filters can be used to find the ROI center. Recall that, in addition to the state prediction, the filters provide a prediction covariance that is a measure of the filter's perceived accuracy in the prediction. This prediction covariance defines a confidence ellipse over space that can be used to determine the size of the ROI in each direction [6, 54, 83]. Mathematically, the ROI is defined as the g-sigma confidence ellipsoid [6, 83] or the locus of spatial points $\mathbf{p}$ such that

$$(\mathbf{p} - \mathbf{A}\hat{\mathbf{x}}(k|k-1))^T \mathbf{A}\mathbf{P}(k|k-1)^{-1}\mathbf{A}^T (\mathbf{p} - \mathbf{A}\hat{\mathbf{x}}(k|k-1)) = g^2. \qquad (8.6)$$

The matrix $\mathbf{A}$, defined in Equation (7.11), is used to extract x and y position estimates from the state vector. The lengths of the semi-axes of the above ellipsoid are g times the square root of the eigenvalues of the covariance matrix. Therefore, this parameter controls the probability that the difference between the true and estimated positions lies within the ellipsoid defined in Equation (8.6). For example, the probability that the error vector is inside the 3-sigma and 4-sigma ellipsoids is 98.89% and 99.97%, respectively.

Denote the number of anchor points inside the ROI at time k as $N(k)$. The localized radio map is now defined as $\mathcal{R}(k) = \{(\mathbf{p}_i, \mathbf{F}(\mathbf{p}_i))\}_{i=1}^{N(k)}$. Subsequent positioning operations will use only anchor points in the ROI.

Recall that in Chapter 7, we discussed the role of feedback in adaptive filtering (recursive filtering). This type of feedback is different than that used in ROI selection. In particular, we distinguish between two types of feedback.

- **Local feedback** This type of feedback is used internally by the adaptive filters to implement the recursive predictor–corrector structure. This allows the filter to adjust the relative weights of the prediction and measurement contributions. This type of feedback is *reactive* in that parameter adjustment is only performed after receiving an observation.
- **Global feedback** This type of feedback connects two distinct modules of the tracking system, namely sensor selection and positioning. This type of feedback is *proactive* in that the system parameters (anchor points included in positioning) are determined prior to receiving an observation.

Caution must be taken when implementing feedback systems as erroneous position predictions affect not only the positioning accuracy at the current time step, but also the anchor point and access point choices for future estimates. This is of special concern in the WLAN tracking problem where RSS observations are known to be noisy. In this light, an outlier detection scheme is needed so that outlier observations can be discarded [54].

8.4   Experimental examples

In this section, we demonstrate the advantages of the sensor selection methods discussed in this chapter using the two experimental data sets. The NI filter is chosen as an illustrative example, though the results can be generalized to other adaptive filters.

8.4.1   Size of ROI

Figures 8.2 and 8.3 depict the performance of the NI filter as a function of the size of the ROI for the stationary and mobile test scenarios for both validation and test sets. In these examples, the size of the ROI is varied by varying the parameter g. Note that as the size of the ROI grows, the performance of the system converges to that of a system with no ROI consideration (no anchor point selection).

As seen from Figure 8.2, the effect of anchor point selection is minimal on the system accuracy when the mobile user remains stationary.

For the mobile user, anchor point selection has a large effect on the performance of the system. This result is expected because the ROI is meant to combat the undesirable variations in RSS that lead to inclusion of erroneous anchor points in position estimation. As such, we expect that the benefits of this methods would be particularly evident in the case of the mobile user, where motion of the device leads to an increased variance in RSS measurements.

A second observation is that the optimal size of the ROI corresponds to $g \approx 4$. Recall that this value corresponds to the case that the error vector is inside the confidence ellipse with probability 99.97%. It is also encouraging to note that positioning accuracy is not highly sensitive to this parameter.

8.4.2   Access point selection

Figure 8.4 shows the effect of access point selection on positioning accuracy for different selection techniques.

As seen from the results, the strongest access point selection method has the poorest performance among the methods discussed in this chapter. This is expected as this technique does not consider the positioning algorithm in its selection strategy. Moreover, as discussed previously, the RSS observations received from the strongest access points are generally associated with the highest variances, increasing the chance of discrepancy between measured values and those stored in the radio map.

The improvements in positioning accuracy due to access point selection are particularly evident in the case of the mobile device. Interestingly, we see that these improvements occur when feedback-based ROI selection is performed (note that the differences in positioning accuracy using different access point selection methods are minimal for large values of the parameter g). This result suggests that access point selection must be carried out over localized spatial regions. This is consistent with the notion that the statistics of RSS values received from an access point vary over space. Therefore,

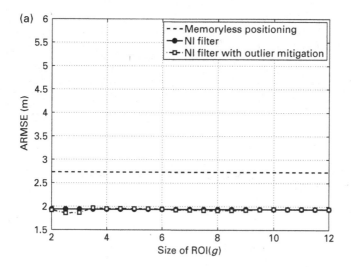

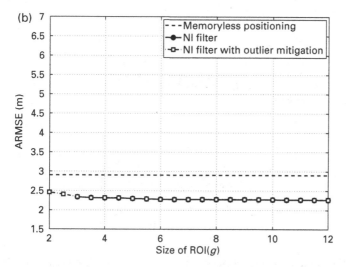

Figure 8.2. Effect of the size of ROI(g) on average root mean square error (ARMSE). (a) Stationary user (validation). (b) Stationary user (test).

the entire region of coverage may not provide an informative platform for optimization of access point selection criteria. This suggests that access point selection criteria should be evaluated over localized spacial regions.

8.4.3 The role of feedback

Tables 8.1 and 8.2 compare the performance of three positioning methods: memoryless positioning using the kernel density estimator (KDE), dynamic positioning using the NI

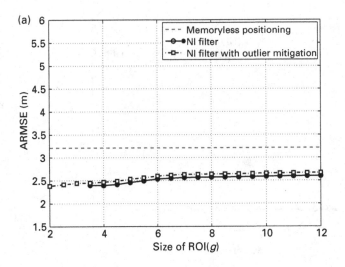

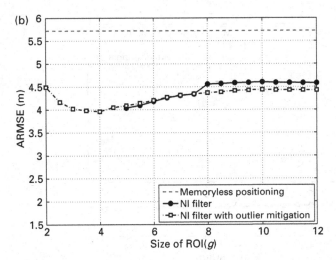

Figure 8.3.   Effect of the size of ROI (g) on average root mean square error (ARMSE). (a) Mobile user (validation). (b) Mobile user (test).

filter, and positioning with anchor point and access point selection for both the stationary and mobile data sets.

For the stationary user, we note that all methods provide similar positioning accuracies. This, however, is not the case for the mobile user. The results of Table 8.2 show that positioning accuracy improves successively as model-based positioning and sensor selection components are added to the system. The most impressive observation is that the combination of adaptive filtering and sensor selection is able to produce positioning accuracies for the mobile user that are comparable to that of the stationary case.

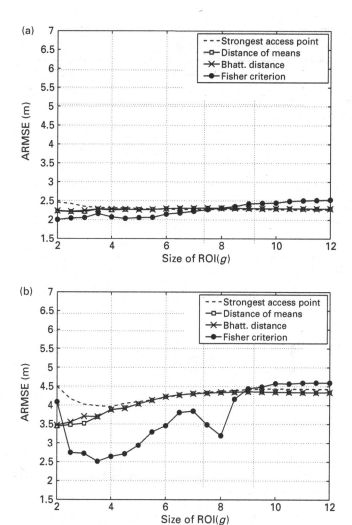

Figure 8.4. Effect of access point selection strategy on average root mean square error (AMRSE) (test data sets). (a) Stationary user. (b) Mobile user.

8.4.4 Tracking examples

Figure 8.5 shows positioning examples for various scenarios. In particular, this figure shows examples for cases where the mobile user is simply moving along a straight path, when the user reverses the direction of uses, when the user stops along their path, and when the user user stops to interact with other people along their path.

As seen, in each case the positioning system is able to track the mobile device successfully.

Table 8.1. Positioning error statistics for memoryless KDE estimator, the NI filter, the anchor point and access point selection methods (stationary user)

Method	ARMSE (m)	Mean (m)	Variance (m^2)
Memoryless KDE	2.90	1.83	4.98
NI filter	2.29	1.66	3.00
NI filter (Fisher + anchor point selection)	2.07	1.80	2.31

Table 8.2. Positioning error statistics for memoryless KDE estimator, the NI filter, the cognitive anchor point and access point selection methods (mobile user)

Method	ARMSE (m)	Mean (m)	Variance (m^2)
Memoryless KDE	5.70	3.94	24.25
NI filter	4.58	3.37	16.68
NI filter (Fisher + anchor point selection)	2.51	1.97	2.96

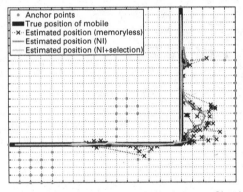
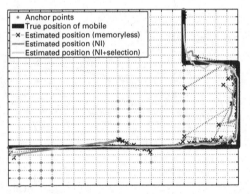

Simple motion

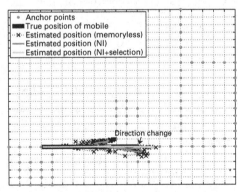
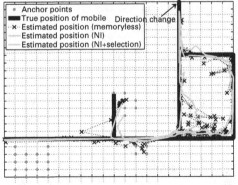

Direction reversal

Figure 8.5.   Example track results.

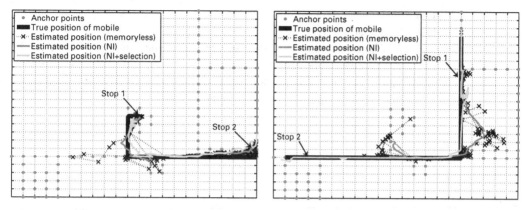

Changing velocity

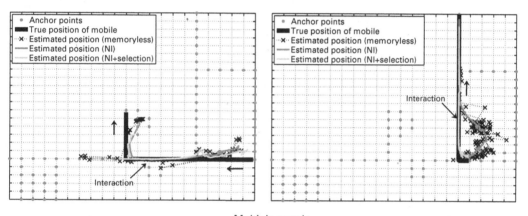

Multiple people

Figure 8.5. (cont.)

8.5 Chapter summary

In this chapter, we discussed sensor selection strategies used to improve the accuracy of positioning and reduce computational complexity. In particular, we focused on techniques for selection of access points and anchor points.

With respect to access point selection, we presented several methods that aim to minimize the redundancy between the selected access points or to maximize their discrimination ability. We saw that these methods can significantly improve positioning accuracy when compared to traditional selection methods relying on the RSS.

In terms of anchor point selection, we discussed methods for localizing positioning operations to a region of interest (ROI) in the coverage area of the positioning system. We noted that the ROI can be derived in a memoryless manner or proactively using covariance estimates from adaptive filters. Experimental results showed that anchor point selection can significantly improve positioning accuracy by combating the undesirable effects of RSS variations over time.

Overall, access point and anchor point selection strategies discussed herein can be considered as adaptive radio scene analysis methods. These techniques localize positioning operations to a subset of the radio scene to mitigate the effects of unpredictable RSS variations and minimize positioning complexity.

9   System design considerations

In the previous chapters, we addressed the problem of position estimation given values of received signal strength (RSS). In this chapter, we focus on the architectural and system design aspects of a positioning solution.

We begin by outlining the five design issues that must be considered in WLAN positioning systems (Section 9.1). We then proceed to discuss architectural details of three functional modules of positioning systems, namely sensing, computation, and storage (Sections 9.2, 9.3, 9.4, and 9.5).

9.1   Design issues

Before proceeding to a discussion of the design of an RSS-based positioning system in detail, let us provide an overview of five key design challenges in these systems.

Limited communication bandwidth

Communication via wireless channels imposes severe limitations on the available bandwidth (5 and 11 Mbps for an IEEE 802.11b wireless LAN [81]). This is due to the intensity drop as a function of the fourth power of the distance, the presence of obstructing objects (shadowing effects), and multipath fading [69]. As a result, RSS-based positioning systems must be designed in such a way as to minimize the number and frequency of wireless communications.

Distributed operation

Because communication is much more expensive than most processing operations, it is highly desirable for the processing to take place in a distributed manner to reduce traffic volume and energy costs [69]. Also, in large environments, access and anchor points are distributed over a large geographical area. Thus, distributed and localized computation of location estimates as dictated by the physical configuration of the environment is often advantageous in reducing computation and communication costs.

Insecure communication channels

Since positioning systems collect and analyze information on people, privacy and protection of the sensor data is of paramount importance. This means that restrictions on the nature of data communicated over insecure channels may be necessary.

Scalability

RSS-based positioning systems may be deployed in large indoor environments such as shopping malls, museums, hospitals, and airports. These systems must be able to sustain a large number of users at a given time. Moreover, the number of anchor and access points generally grows as the area of the environment. Scalability with respect to the number of users and area of the environment is therefore a chief concern in these systems.

Unreliable observations

Classical positioning methods generally rely on a continuous stream of RSS readings. However, factors such as interference, non-line-of-sight (NLOS) propagation, and mobility of users may greatly influence the reliability of data transmitted over a wireless channel. Furthermore, the limitations of battery operated devices, as well as communication delays among devices, may lead to intermittent and asynchronous observations. A positioning system must be designed in such a way that minimizes the effect of such disruptions on the positioning service provided to the user. The adaptive and proactive systems discussed in previous chapters are a first step in this direction.

9.2   Functional units

In this chapter, we shall discuss the architectural details of three functional units of a positioning system, namely sensing, computation, and storage modules (Figure 9.1). The sensing unit encompasses all operations related to obtaining RSS measurements and concerns interactions and exchanges between the access points and mobile devices in the network. The computation module is responsible for transforming RSS measurements into position estimates. Finally, the storage unit manages storage and access to radio map information.

As shown in Figure 9.1, the three functional components of a positioning system may interact with each other. For example, the sensing unit may provide the storage unit with new information to update the radio map. Similarly, the sensing and computation units can interact either when new RSS information is forwarded to the positioning engine, or when the positioning system guides sensing operations.

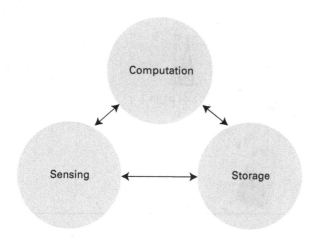

Figure 9.1.   Functional components of an RSS-based WLAN positioning system.

9.3   Sensing

Sensing in the context of RSS-based WLAN positioning is the measurement of RSS values. As shown in Figure 9.2, two sensing paradigms can be employed. In the first paradigm (Figure 9.2(a)), the mobile device sends out probe requests to the access points and measures the strength of the response beacons. In the second paradigm (Figure 9.2(b)), access points measure the strength of the signal received from the mobile device.

The first paradigm is often the one employed in positioning systems. This method is advantageous in terms of scalability of the system. In fact, a positioning system may have a large number of subscribers at a time. Allowing the sensing to be carried out in a distributed manner by each device means that the sensing requirements of the system remain constant as the number of subscribers increases. Moreover, in this paradigm minimal communication over the network is required. In particular, the only communication operations needed are transmission of probe requests by the mobile client to each access point and probe responses by the access points to the mobile device (whereas the second paradigm requires further transmission of RSS values to the positioning engine).

9.4   Computation

The computation of a position estimate using the methods discussed in previous sections requires two types of information to be fused: (1) information from multiple access points and (2) information from multiple anchor points. Conforming to the terminology used in data fusion literature, we refer to both anchor points and access points as "sensors." Three architectures for data fusion are available, namely centralized fusion, decentralized

(a)

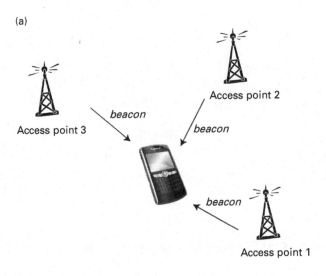

(b)

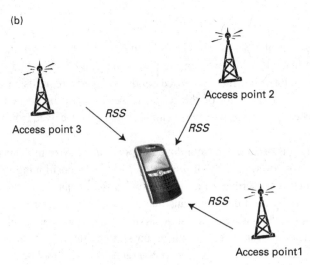

Figure 9.2.   Overview of sensing paradigms in WLAN positioning. (a) Sensing operations performed by the infrastructure. (b) Sensing operations performed by the mobile device.

fusion, and hierarchical fusion architectures. The details of each of these architectures are discussed below.

9.4.1   Centralized architecture

In a centralized fusion architecture, all sensor information (in this case, readings from access points and training values at anchor points) are available to a central data fuser. This central fuser performs all processing and data fusion steps. An example of such an architecture is shown in Figure 9.3(a).

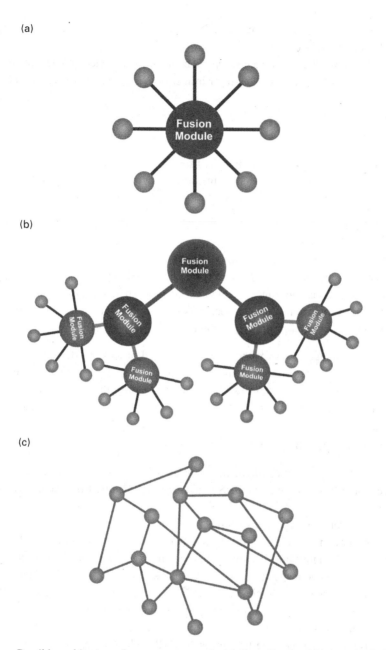

Figure 9.3. Possible architectures for a sensor network. (a) Centralized architecture for fusion. (b) Hierarchical architecture for fusion. (c) Decentralized architecture for fusion.

Centralized fusion systems have access to all observations (e.g. access point and anchor point information) when forming the position estimate [94]. As such, they can detect outlier information and exclude this information during fusion [61]. These systems generally perform best among the various architectures in terms of accuracy. These advantages come at the cost of increased processing demand on the central processor (i.e. the mobile device).

In our context, the centralized fuser can be the mobile device or a dedicated positioning engine, resulting in client-based or infrastructure-based designs as discussed in the following.

Client-based systems

This architecture is almost unanimously used in all positioning systems discussed in the literature. In this type of architecture, the mobile device performs all processing and data fusion steps. The client-based architecture offers three important benefits.

(1) A client-based architecture promotes security of position information, as this information is not transmitted over insecure links or stored anywhere on the network.
(2) A client-based architecture provides great advantages in terms of scalability of the system to accommodate a large number of users, as the computational burden on the system remains constant as the number of subscribers increases.
(3) A client-based architecture respects privacy of the mobile client since the only entity with knowledge of position information is the mobile device.

These advantages come at the cost of increased processing demand on the mobile client. This is of special concern in WLAN positioning as the limited processing and battery power of mobile devices may restrict not only the frequency of positioning computations, but also the complexity of positioning algorithms employed.

Infrastructure-based systems

In this type of architecture, a dedicated positioning engine is responsible for computation of location information based on RSS values received either from the mobile device or the access points.

Infrastructure-based designs do not suffer from the previously discussed computational power and battery limitations. However, the need to communicate RSS readings to the central engine translates to an inefficient use of network bandwidth. This is a disadvantage in two respects. First, communication costs are generally higher than most processing operations. As such, requiring the mobile device to transmit RSS information continuously to the positioning engine imposes a significant power burden on battery-operated mobile devices. The second disadvantage of this architecture relates to security concerns connected with the transmission of RSS values over insecure wireless links which are susceptible to eavesdropping and malicious attacks [53].

9.4.2   Hierarchical architecture

As depicted in Figure 9.3(b), a hierarchical architecture allows for the fusion process to take place at various levels in the system.

The fingerprinting-based methods discussed in Chapter 6 are especially suitable for a hierarchical architecture. In these designs each anchor point determines its weight using the RSS signature of that particular point (note that in doing so, information from multiple access points is fused). The weights computed by each anchor point are then used by a positioning engine to determine the final position estimate.

Another way to implement a hierarchical architecture is to require each access point to form a position estimate by fusing information from multiple anchor points. These estimates are then fused to arrive at a final position. This method is employed in [53].

In a hierarchical architecture, computations may take place in a hierarchical manner on one computational unit (for example, the mobile device) as is the case in the previous examples. It is also possible that computations are distributed across sensors that are physically distinct. In the latter case, the hierarchical architecture overcomes the limitation of a centralized architecture in terms of computational complexity. However, this architecture suffers from the limitations of the infrastructure-based system. Moreover, two-way communication between the central positioning engine and the local anchor/access points may be required [61].

9.4.3   Decentralized architecture

In a decentralized architecture, each sensor node (either access point or anchor point) has the processing power to form estimates based on local observations. Furthermore, each sensor receives data from other sensors connected to it and performs fusion to generate a global estimate. There are many advantages of using multiple sensors, two of which are [66]:

- redundancy – failure of one sensor does not result in failure of the entire system;
- reliability – erroneous readings from one malfunctioning sensor do not necessarily degrade the performance of the system.

If the sensors are distinct computational entities, communication between sensors is required. This adds a significant burden on the wireless infrastructure. Distributed architectures in WLAN positioning have not been explored to date, though techniques developed for distributed positioning in wireless sensor networks may be applied to WLAN positioning.

9.5   Storage

The third functional module in WLAN positioning is responsible for storage of the radio map. Due to the critical role of the radio map during positioning, this radio map must be secure from malicious attacks and easily maintainable. In infrastructure-based positioning systems, the map can be stored on the positioning engine. In client-based systems, however, the map must be accessible to all subscribers. As such, it is generally assumed that the map is stored on a central server and transmitted over wireless links to mobile clients. The following sections discuss design concerns related to access, security, and maintainability of the map.

9.5.1   Access

The radio map provides the positioning engine with essential information for computation of the position. For a client-based centralized architecture, the radio map must be available to all subscribers to the system. Several design paradigms may be considered for providing access to the map. In the most basic design, the radio map is transmitted to the mobile client in its entirety prior to the start of positioning. This transmission can be performed over wired or wireless links. Clearly, the size of the radio map grows with the size of the environment (number of anchor points) and number of access points. As such, in large environments, such as shopping malls, transmission of the entire map may require significant bandwidth and impose a storage burden on the mobile device.

Another design may transmit only relevant portions of the radio map to the mobile device depending on its position [47]. This design provides obvious advantages in transmission and storage complexity. At the same time, implementation of this paradigm requires predictive knowledge of the future position of the mobile so that the appropriate section of the maps is available to the mobile device as needed. As such, this design is especially suitable for use with dynamic positioning systems.

9.5.2   Security

Positioning systems generally rely on data transmitted over insecure wireless links (for example, in passive sensing of RSS from access points). These data are susceptible to malicious security attacks [53]. The objective of such attacks may be to disrupt the positioning service or to obtain unauthorized user position information.

Security attacks may affect the off-line training phase or the on-line positioning. During off-line training, an adversary can impersonate access points (Sybil attack), compromising the verity of the collected RSS information. Similarly, an adversary may jam the signal from one or more access points or attenuate or amplify these signals. To remedy impersonation attacks, authentication beacon nodes may be used [59]. Corruptions to training RSS values can be partially mitigated through collection of data over multiple time windows. Moreover, validation and attack detection schemes [13] can be applied to the data once the training data are collected.

The second vulnerability of WLAN positioning systems to malicious attacks comes from their reliance on the radio map information. In particular, an adversary can disrupt positioning services through malicious manipulation of the radio map information. Such manipulation can occur at the positioning server end where the radio map is stored or during transmission of the fingerprint data to the mobile client. These vulnerabilities can be mitigated by effectively securing the positioning server and by using encryption techniques to protect information transmitted over insecure links.

RSS-based attacks can also be carried out on the observed radio signal during the on-line operation of the system. In such attacks an adversary alters RSS readings through impersonation, jamming, and attenuation or amplification [58]. Such attacks can be combatted through implementation of attack detection algorithms [13], or by increasing system resiliency to attacks through increased redundancy (i.e. increased number of access points) [58]. It is also possible to use a set of static and hidden base-stations for secure positioning [11]. Finally, resiliency to RSS attacks can be achieved through distributed positioning whereby information from access points is selectively used depending on the estimated reliability of each access point [53].

9.5.3   Maintainability

As previously mentioned, the radio propagation environment is affected by shadowing and multipath effects caused by absorption and reflection of radio waves. As such, changes in the layout and location of obstructing objects (for example, furniture) significantly affect the spatial and temporal distribution of RSS. In addition, WLAN infrastructures are highly dynamic, as access points can easily be moved or discarded. These changes may degrade the performance of positioning systems if they result in significant deviations between the RSS signatures stored in the radio map and the RSS values measured during the on-line operation of the system.

To mitigate the adverse effects of changes in the propagation environment and access points, the radio map must be updated regularly. In the simplest case, the entire map is periodically updated through collection of new training data. This method, however, suffers from a relatively high cost associated with laborious data collection in large environments. More efficient maintenance strategies include selective updating of a portion of the radio map based on a predefined fidelity criterion. Another technique is the on-line determination of the radio map by using a set of reference RSS receivers to sense the environment continuously and adapt the entire radio map accordingly through interpolation and regression analysis.

9.6   Chapter summary

In this chapter we discussed key challenges in designing WLAN-based positioning systems. Furthermore, we outlined architectural considerations for designing the sensing, computation, and storage modules in these systems.

The key considerations in designing the sensing module in WLAN systems are communication cost and security. In this light, performing sensing operations on the mobile client is preferred to carrying out these operations on access points.

For the design of the computational unit, we reviewed centralized, hierarchical, and distributed architectures. Considering the cost of communication in WLAN networks, most existing systems carry out computational operations centrally on the mobile device.

Finally, we discussed issues related to access, maintenance, and security of the radio map. Currently, most systems assume that the radio map is transmitted to the mobile client either at the start of positioning or on an on-demand basis. Interesting future research directions include dynamic and automatic updating of the radio map.

10   The road ahead

Since the first WLAN-positioning system was introduced in 2000 [4], rapid advances in signal processing methods have been made in this area. A decade later, fundamental positioning techniques have matured significantly, allowing these systems to offer highly accurate positioning estimates with accuracies on the order of several meters in both indoor and outdoor environments.

The previous chapters have reviewed the fundamental techniques used in WLAN positioning. In this chapter, we look ahead to opportunities and challenges remaining to be addressed in this area.

10.1   Highlights

The focus of this book has been WLAN-based positioning. Chapters 1 to 4 discussed the history, applications, and various positioning systems to motivate these systems. In particular, the development of these systems is motivated by the need for accurate, reliable, and cost-efficient positioning solutions to enable the delivery of location-based services (LBS).

The second part of the book was dedicated to fundamental signal processing concepts in these systems. In Chapter 5, we saw that the unpredictability of radio signal features poses a significant challenge to the development of accurate and reliable WLAN systems. In Chapter 6, we discussed a number of non-parametric techniques that can be used to model these radio signals using training samples collected at a set of anchor points with known locations. In addition to their effectiveness as modeling tools, these non-parametric techniques also allowed the estimation of a measure of uncertainty associated with position estimates. The availability of this uncertainty measure is essential if the positioning information is to be used further in systems that offer LBS in a consistent and reliable manner.

To improve the accuracy and reliability of positioning further, we discussed a number of data fusion techniques in Chapter 7. In particular, we considered adaptive filtering techniques that fuse knowledge of motion dynamics with WLAN measurements to enhance accuracy and reliability of the positioning system. Finally, we presented a number of issues related to architecture and design of WLAN systems in Chapter 9.

10.2 Directions for future research

The late 1990s and early years of the twenty-first century have witnessed significant milestones in the development of accurate and reliable WLAN positioning systems. However, several challenges remain to be addressed.

10.2.1 Automatic modeling tools

Most existing WLAN positioning systems employ the fingerprinting technique, which uses training measurements to model the RSS–position relationship. One of the main limitations of WLAN systems relying on fingerprinting is the sensitivity of the training location fingerprints to environmental changes and device characteristics. Furthermore, the laborious nature of fingerprint collection hinders the scalability of the proposed technique to large environments.

An interesting solution for overcoming these limitations is the use of dynamically built radio maps based on realtime sensing of the environment. Recent literature has considered using scanning nodes that provide continuous measurements of the environment. Another approach would be to use position estimates and the corresponding RSS observations from various mobile estimates to update the radio map gradually. Further study of these methods can significantly benefit WLAN positioning by reducing the labor cost and improving the accuracy of the radio map.

10.2.2 Cooperative positioning

Positioning techniques discussed in this book consider a mobile device in isolation. This approach does not reflect the reality, in that mobile devices are generally in environments where other users access positioning services, potentially from other positioning systems. A more general model of the positioning environment can lead to the design of *cooperative* positioning systems. In these systems, positioning can benefit from cooperation on four levels as discussed below.

- **Cooperation among users** The increasing number of subscribers to positioning services motivates the development of cooperative positioning solutions. This type of cooperation can be used in conjunction with the designs proposed here to guide the proactive selection of sensing parameters as well as deployment of necessary resources. Further, such collaboration may be used to relay channel information from one user to another. For example, dynamically chosen mobile clients may serve as anchor points to update training information for other users. In addition, comparison of RSS signals between users may be used further to model signal variability over space.
- **Cooperation among sensors** The accuracy of RSS-based positioning is fundamentally limited due to the nature of the radio channel. As such, sensors of other, complementary modalities can be used to enhance positioning accuracy and reliability. Fusion of radio and video is especially of interest because of the complementarity

of the two modalities with respect to noise process, cost, computational complexity, and accuracy.

- **Cooperation among users and systems** An interesting avenue for improving accuracy and reliability of positioning systems may employ *interactivity*. Such a design may, for example, employ user feedback on accuracy to re-initialize system parameters, determine mobility models, or determine the need to update system parameters.

- **Cooperation among positioning systems** In order to provide seamless and transparent LBS to mobile users, several positioning systems may need to cooperate (for example, when a user moves from an outdoor environment to an indoor environment). An important issue to consider in this regard is hand-off algorithms between location systems. In addition, such schemes necessitate the development of standard data formats that can be used by positioning systems to communicate.

10.2.3   Proactive systems

Most existing LBS systems are *reactive* systems. These systems provide personalized information based on the current position of a mobile user only. An interesting direction for future research is to consider the use of proactive or predictive positioning information. These systems can anticipate and respond to user needs in a context-dependent manner. The adaptive filtering techniques discussed in this book are especially suitable for the development of such systems as they provide predictions of future user positions.

10.2.4   Applications

Positioning technology is only just coming of age. However, the past years have seen a drastic change in the economic landscape of LBS. We anticipate that the ever-increasing market penetration of mobile devices equipped with positioning technologies will lead to new generations of innovative applications where location information can be used to enhance the quality of users' experience, and to add value to existing services offered by wireless providers. This necessitates the development of new business models that represent the opportunities offered by positioning systems. Moreover, new standards for management of location information must be developed to facilitate information sharing between multiple applications and positioning systems.

10.2.5   New challenges

This book has mainly discussed the potential of location information and what can be achieved with positioning systems. We would like now to shift our focus to the ethical implications of positioning. As mentioned in the opening chapter of the book, location computing is associated with various ethical issues related to privacy and consent. It is important that future research focuses on ethical aspects associated with these technologies as much as it emphasizes technological advancements. Moreover, significant strides must still be made to develop policy and regulations associated with location information. In this light, we would like to conclude this book by posing the following question: What *should* be done with location information?

References

[1] E. Aarts and B. Eggen. *Ambient Intelligence in HomeLab*. Eindhoven: Neroc Publishers, 2002.

[2] G. Antonini. A discrete choice modeling framework for pedestrian walking behavior with application to human tracking in video sequences. Ph.D. thesis, École Polytechnique Fèdèrale de Lausanne, 2006.

[3] M. S. Arulampalam, S. Maskell, N. Gordon, and T. Clapp. A tutorial on particle filters for online nonlinear/non-gaussian bayesian tracking. *IEEE Transactions on Signal Processing*, 50(2):174–188, 2002.

[4] P. Bahl and V. N. Padmanabhan. RADAR: an in-building RF-based user location and tracking system. In *Proc. IEEE Infocom*, Tel Aviv, 2000, vol. 2, pp. 775–784.

[5] R. M. Balan. Q-markov random probability measures and their posterior distributions. *Stochastic Processes and their Applications*, 109(2):295–316, 2004.

[6] Y. Bar-Shalom, X.-R. Li, and T. Kirubarajan. *Estimation with Applications to Tracking and Navigation*. New York: Wiley, 2001.

[7] D. Basu. *Statistical Information and Likelihood: A Collection of Critical Essays*. New York: Springer-Verlag, 1988.

[8] A. R. Beresford and F. Stajano. Location privacy in pervasive computing. *IEEE Pervasive Computing*, 2(1):46–55, 2003.

[9] D. Bosq. *Nonparametric Statistics for Stochastic Processes: Estimation and Prediction*. Lecture Notes in Statistics 110. New York: Springer, 1998.

[10] M. Brunato and R. Battiti. Statistical learning theory for location fingerprinting in wireless LANs. *Computer Networks*, 47(6):825–845, 2005.

[11] S. Capkun, M. Cagalj, and M. Srivastava. Secure localization with hidden and mobile base stations. In *Proc. 25th IEEE Int. Conf. Computer Communications (INFOCOM)*, Barcelona, 2006, p. 1–10.

[12] A. F. Cattoni, A. Dore, and C. S. Regazzoni. Video-radio fusion approach for target tracking in smart spaces. In *Proc. IEEE Int. Conf. Information Fusion*, Quebec, 2007.

[13] Y. Chen, W. Trappe, and R. P. Martin. ADLS: Attack detection for wireless localization using least squares. In *Proc. 5th Annu. IEEE Int. Conf. Pervasive Computing and Communications Workshops (PerComW'07)*, 2007, pp. 610–613.

[14] Y. Chen, Q. Yang, J. Yin, and X. Chai. Power-efficient access-point selection for indoor location estimation. *IEEE Transactions on Knowledge and Data Engineering*, 18(7):877–888, 2006.

[15] Computer Science and Artificial Intelligence Laboratory, Massachusetts Institute of Technology. MIT: Project Oxygen, 2004.

[16] P. M. Djuric, J. H. Kotecha, J. Zhang, Y. Huang, T. Ghirmai, and M. F. Bugallo Joaquin Miguez. Particle filtering. *IEEE Signal Processing Magazine*, 20(5):19–38, 2003.

[17] A. Dore, A. F. Cattoni, and C. S. Regazzoni. A particle filter based fusion framework for video-radio tracking in smart spaces. In *Proceedings IEEE Conf. Advanced Video and Signal Based Surveillance (AVSS)*, 2007, pp. 99–104.

[18] A. Doucet, S. Godsill, and C. Andrieu. On sequential Monte Carlo sampling methods for bayesian filtering. *Statistics and Computing*, 10(3):197–208, 2000.

[19] R. O. Duda, P. E. Hart, and D. G. Stork. *Pattern Classification*, 2nd edn. New York: Wiley, 2001.

[20] P. Enge and P. Misra. Special issue on global positioning system. *Proceedings of the IEEE*, 87(1):3–15, 1999.

[21] S. H. Fang and T. N. Lin. Indoor location system based on discriminant-adaptive neural network in IEEE 802.11 environments. *IEEE Transactions on Neural Networks*, 19(11):1973–1978, 2008.

[22] S. H. Fang, T. N. Lin, and P. C. Lin. Location fingerprinting in a decorrelated space. *IEEE Transactions on Knowledge and Data Engineering*, pp. 685–691, 2007.

[23] Federal Communications Commission (FCC). Report and order and further notice of proposed rulemaking in the matter of revision of the commission's rules to ensure compatibility with enhanced 911 emergency calling systems, 1996.

[24] J. Figueiras and S. Frattasi. *Mobile Positioning and Tracking*. Wiley Online Library, 2010.

[25] Georgia Institute of Technology. Aware Home Research Initiative, 2004.

[26] I. A. Getting. The global positioning system. *IEEE Spectrum*, 30(12):36–47, 1993.

[27] A. Goldsmith. *Wireless Communications*. Cambridge: Cambridge University Press, 2005.

[28] Adam Greenfield. *Everyware: The Dawning Age of Ubiquitous Computing*. Berkeley, CA: New Riders, 2006.

[29] F. Gustafsson, F. Gunnarsson, N. Bergman, U. Forssell, J. Janssona, R. Karlsson, and P.-J. Nordlund. Particle filters for positioning, navigation, and tracking. *IEEE Transactions on Signal Processing*, 50(2):425–437, 2002.

[30] I. Guvenc, C. T. Abdallah, R. Jordan, and O. Dedeoglu. Enhancements to RSS based indoor tracking systems using Kalman filters. In *Proc. Int. Signal Processing Conference and Global Signal Processing Expo*, Dallas, TX, 2003.

[31] A. Haeberlen, E. Flannery, A. M. Ladd, A. Rudys, D. S. Wallach, and L. E. Kavraki. Practical robust localization over large-scale 802.11 wireless networks. In *Proc. Tenth ACM Int. Conf. Mobile Computing and Networking (MOBICOM)*, Philadelphia, PA, 2004, pp. 70–84.

[32] A. Hampapur, L. Brown, J. Connell, A. Ekin, N. Haas, M. Lu, H. Merkl, and S. Pankanti. Smart video surveillance: exploring the concept of multiscale spatiotemporal tracking. *Signal Processing Magazine, IEEE*, 22(2):38–51, 2005.

[33] S. Haykin. *Adaptive Filter Theory*. Englewood Cliffs, NJ: Prentice Hall, 2002.

[34] M. Hazas, J. Scott, and J. Krumm. Location-aware computing comes of age. *IEEE Computer*, 37(2):95–97, 2004.

[35] J. Hightower, B. Brumitt, and G. Borriello. The location stack: a layered model for location in ubiquitous computing. In *Proc. 4th IEEE Workshop Mobile Computing Systems and Applications*, New York, 2002, pp. 22–28.

[36] Y. Ho and R. Lee. A Bayesian approach to problems in stochastic estimation and control. *IEEE Transactions on Automatic Control*, 9(4):333–339, 1964.

[37] W. Hu, T. Tan, L. Wang, and S. Maybank. A survey on visual surveillance of object motion and behaviors. *IEEE Transactions on Systems, Man and Cybernetics, Part C*, 34(3):334–352, August 2004.

[38] IEEE. IEEE Standard for information technology – Telecommunications and information exchange between systems – local and metropolitan area networks – specific requirements – Part 11: wireless LAN medium access control (MAC) and physical layer (PHY) specifications, 2007.

[39] A. J. Izenman. Recent developments in nonparametric density estimation. *Journal of the American Statistical Association*, 86:205–224, 1991.

[40] Y. Jie, Y. Qiang, and N. Lionel. Learning adaptive temporal radio maps for signal-strength-based location estimation. *IEEE Transactions on Mobile Computing*, 7(7): 869–883, 2008.

[41] K. Kaemarungsi. Distribution of WLAN received signal strength indication for indoor location determination. In *Proc. First IEEE Int. Symp. Wireless Pervasive Computing*, Phuket, 2006.

[42] K. Kaemarungsi and P. Krishnamurthy. Modeling of indoor positioning systems based on location fingerprinting. In *Proc. 23rd Annu. Joint Conf. IEEE Computer and Communications Societies (INFOCOM)*, Hong Kong, 2004, vol. 2, pp. 1012–1022.

[43] K. Kaemarungsi and P. Krishnamurthy. Properties of indoor received signal strength for WLAN location fingerprinting. In *Proc. 1st Annu. Int. Conf. Mobile and Ubiquitous Systems: Networking and Services (MOBIQUITOUS)*, Boston, MA, 2004, pp. 14–23.

[44] T. Kailath, A. H. Sayed, and B. Hassibi. *Linear Estimation*. Upper Saddle River, N. J.: Prentice Hall, 2000.

[45] R. E. Kalman. A new approach to linear filtering and prediction problems. *Journal of Basic Engineering*, 82(1):35–45, 1960.

[46] M. B. Kjærgaard. A taxonomy for radio location fingerprinting. *Lecture Notes in Computer Science*, 4718:139–156, 2007.

[47] M. B. Kjærgaard, G. Treu, P. Ruppel, and A. Küpper. Efficient indoor proximity and separation detection for location fingerprinting. In *Proc. 1st Int. Conf. MOBILe Wireless MiddleWARE, Operating Systems, and Applications*, ICST (Institute for Computer Sciences, Social Informatics and Telecommunications Engineering), 2008, pp. 1–8.

[48] M. Kjaergaard and C. Munk. Solving RSS client differences by hyperbolic location fingerprinting. In *Proc. IEEE Int. Conf. Pervasive Computing and Communications*, Alexandria, 2008.

[49] P. Krishnan, A. S. Krishnakumar, W.-H. Ju, C. Mallows, and S. Ganu. A system for LEASE: Location estimation assisted by stationary emitters for indoor RF wireless networks. In *Proc. IEEE Infocom*, Hong Kong, 2004, vol. 2, pp. 1001–1011.

[50] A. Küpper. *Location-Based Services: Fundamentals and Operation*. Chichester: John Wiley and Sons, 2005.

[51] A. Kushki, K. N. Plataniotis, and A. N. Venetsanopoulos. Location tracking in wireless local area networks with adaptive radio maps. In *Proc. IEEE Int. Conf. Acoustics, Speech, and Signal Processing (ICASSP)*, Toulouse, 2006, vol. 5, pp. 741–744.

[52] A. Kushki, K. N. Plataniotis, and A. N. Venetsanopoulos. Kernel-based positioning in wireless local area networks. *IEEE Transactions on Mobile Computing*, 6(6):689–705, 2007.

[53] A. Kushki, K.N. Plataniotis, and A.N. Venetsanopoulos. Sensor selection for mitigation of RSS-based attacks in wireless local area network positioning. In *Proc. IEEE Int. Conf. Acoustics, Speech and Signal Processing (ICASSP)*, Las Vegas, NV, 2008, pp. 2065–2068.

[54] A. Kushki, K. N. Plataniotis, and A. N. Venetsanopoulos. Cognitive dynamic tracking for indoor wireless local area networks. *IEEE Transactions on Mobile Computing*, 9(3):390–404, 2010.

[55] A. M. Ladd, K. E. Bekris, A. Rudys, G. Marceau, L. E. Kavraki, and D. S. Wallach. Robotics-based location sensing using wireless ethernet. In *Proc. 8th ACM Int. Conf. Mobile Computing and Networking (MobiCom)*, Atlanta, GA, 2002, pp. 227–238.

[56] A. LaMarca and E. de Lara. Location systems: An introduction to the technology behind location awareness. *Synthesis Lectures on Mobile and Pervasive Computing*, 3(1), 2008.

[57] X. Li and K. Pahlavan. Super-resolution TOA estimation with diversity for indoor geolocation. *IEEE Transactions on Wireless Communications*, 3(1):224–234, 2004.

[58] Z. Li, W. Trappe, Y. Zhang, and B. Nath. Robust statistical methods for securing wireless localization in sensor networks. In *Proc. 4th Int. Symp. Information Processing in Sensor Networks (IPSN)*, Los Angeles, CA, 2005, pp. 91–98.

[59] D. Liu, P. Ning, and W. K. Du. Attack-resistant location estimation in sensor networks. In *Proc. 4th Int. Symp. Information Processing in Sensor Networks (IPSN)*, Los Angeles, CA, 2005, pp. 99–106.

[60] Y. Liu and Z. Yang. *Location, Localization, and Localizability*. New York: Springer, 2010.

[61] J. Manyika and H. F. Durrant-Whyte. *Data Fusion and Sensor Management: A Decentralized Information-Theoretic Approach*. New York: Ellis Horwood, 1994.

[62] M. McGuire and K. N. Plataniotis. Dynamic model-based filtering for mobile terminal location estimation. *IEEE Transactions on Vehicular Technology*, 52(4):1012–1031, 2003.

[63] M. McGuire, K. N. Plataniotis, and A. N. Venetsanopoulos. Location of mobile terminals using time measurements and survey points. *IEEE Transactions on Vehicular Technology*, 52(4):999–1011, 2003.

[64] M. McGuire, K. N. Plataniotis, and A. N. Venetsanopoulos. Data fusion of power and time measurements for mobile terminal location. *IEEE Transactions on Mobile Computing*, 4(2):142–153, 2005.

[65] T. K. Moon and W. C. Stirling. *Mathematical Methods and Algorithms for Signal Processing*. Upper Saddle River, NJ: Prentice Hall, 2000.

[66] A. G. O. Mutambara. *Decentralized Estimation and Control for Multisensor Systems*. Boca Raton, FL: CRC Press, 1998.

[67] P. Perez, J. Vermaak, and A. Blake. Data fusion for visual tracking with particles. *Proceedings of the IEEE*, 92(3):495–513, 2004.

[68] K. N. Plataniotis, D. Androutsos, S. Vinayagamoorthy, and A. N. Venetsanopoulos. Color image processing using adaptive multichannel filters. *IEEE Transactions on Image Processing*, 6(7):933–949, 1997.

[69] G. J. Pottie and W. J. Kaiser. Wireless integrated network sensors. *Communications of the ACM*, 43(5):51–58, 2000.

[70] N. B. Priyantha, A. Chakraborty, and H. Balakrishnan. The cricket location-support system. In *Proc. 6th Annu. Int. Conf. Mobile Computing and Networking (MobiCom)*, Boston, MA, 2000, pp. 32–43.

[71] O. Rashid, P. Coulton, and R. Edwards. Providing location based information/advertising for existing mobile phone users. *Personal and Ubiquitous Computing*, 12(1):3–10, 2008.

[72] B. Ristic, S. Arulampalam, and N. Gordon. *Beyond the Kalman Filter: Particle Filters for Tracking Applications*. Boston, MA: Artech House, 2004.

[73] T. Roos, P. Myllymäki, H. Tirri, P. Misikangas, and J. Sievänen. A probabilistic approach to WLAN user location estimation. *International Journal of Wireless Information Networks*, 9(3):155–164, 2002.

[74] N. Samama. *Global Positioning: Technologies and Performance*. New York: Wiley-Interscience, 2008.

[75] D. W. Scott. *Multivariate Density Estimation*. New York: John Wiley and Sons, 1992.

[76] N. K. Sharma. A weighted center of mass based trilateration approach for locating wireless devices in indoor environment. In *Proc. 4th ACM Int. Workshop Mobility Management and Wireless Access*, Torremolinos, 2006, pp. 112–115.

[77] J. Shawe-Taylor and N. Cristianini. *Kernel Methods for Pattern Analysis*. Cambridge: Cambridge University Press, 2004.

[78] B. W. Silverman. *Density Estimation for Statistics and Data Analysis*. London: Chapman and Hall, 1986.

[79] R. Singh, L. Macchi, C. Regazzoni, and K. Plataniotis. A statistical modelling based location determination method using fusion in WLAN. In *Proc. Int. Workshop Wireless Ad-hoc Networks*, 2005.

[80] S. Spiekermann. General aspects of location-based services. In J. Schiller and A. Voisard, ed., *Location-Based Services*. San Francisco, CA: Morgan Kaufmann Publishers, 2004.

[81] W. Stallings. IEEE 802.11: moving closer to practical wireless LANs. *IT Professional*, 3(3):17–23, 2001.

[82] Guolin Sun, Jie Chen, Wei Guo, and K.J. Ray Liu. Signal processing techniques in network-aided positioning: a survey of state-of-the-art positioning designs. *IEEE Signal Processing Magazine*, 22(4):12–23, 2005.

[83] H. L. Van Trees. *Detection, Estimation, and Modulation Theory*. New York: John Wiley and Sons, 2001.

[84] D. Tse and P. Viswanath. *Fundamentals of Wireless Communication*. Cambridge: Cambridge University Press, 2004.

[85] J. B. Y. Tsui. *Fundamentals of Global Positioning System Receivers*. Wiley Online Library, 2000.

[86] L. Wang, W. Hu, and T. Tan. Recent developments in human motion analysis. *Pattern Recognition*, 36(3):585–601, 2003.

[87] R. Want, A. Hopper, V. Falcao, and J. Gibbons. The active badge location system. *ACM Transactions on Information Systems (TOIS)*, 10(1):91–102, 1992.

[88] B.-F. Wu, C.-L. Jen, and K.-C. Chang. Neural fuzzy based indoor localization by Kalman filtering with propagation channel modeling. In *Proc. IEEE Int. Conf. Systems, Man and Cybernetics*, Montreal, 2007, pp. 812–817.

[89] Z. Xiang, S. Song, J. Chen, H. Wang, J. Huang, and X. Gao. A wireless LAN-based indoor positioning technology. *IBM Journal of Research and Development*, 48(5/6):617–626, 2004.

[90] M. A. Youssef, A. Agrawala, and A. Udaya Shankar. WLAN location determination via clustering and probability distributions. In *Proc. 1st IEEE Int. Conf. Pervasive Computing and Communications*, Fort Worth, TX, 2003, pp. 143–150.

[91] M. Youssef and A. Agrawala. The Horus WLAN location determination system. In *Proc. 3rd Int. Conf. Mobile Systems, Applications, and Services*, Seattle, WA, 2005, pp. 205–218.

[92] M. Youssef and A. K. Agrawala. Continuous space estimation for WLAN location determination systems. In *IEEE 13th Int. Conf. Computer Communications and Networks*, Chicago, IL, 2004.

[93] M. Youssef and A. K. Agrawala. Handling samples correlation in the Horus system. In *Proc. IEEE Infocom*, Hong Kong, 2004.

[94] Y. Zhu. *Multisensor Decision and Estimation Fusion*. Boston, MA: Kluwer Academic Publishers, 2003.

Index

Printed in the United States
by Baker & Taylor Publisher Services